THESE HAVE ALSO WORKED FOR US AT A

This pamphlet has been produced by a group of members living in and around Manchester who wanted to develop classroom ideas at A level.

They were:
- Val Aspin (Stand Sixth Form College)
- Barbara Binns (Shell Centre/J.M.B.)
- Kathleen Cross (Accrington and Rossendale College)
- Gillian Hatch (Manchester Polytechnic)
- Anne Haworth (Stand Sixth Form College)
- Catherine Mottram (Parrs Wood High School, Manchester)
- Ann Shaw (Bishops High School, Chester)
- David Wilson (Reddish Vale School, Stockport)

The group also wishes to acknowledge ideas contributed by:

- Ian Clough (Accrington and Rossendale College)
- Peter Jones (Bilborough College, Nottingham)
- J M H Wright (Unsworth High School, Bury)

CONTENTS

Introduction   5

Chapter 1   Graphs and Functions   7

Chapter 2   Trigonometry   41

Chapter 3   Vectors   53

Chapter 4   Complex Numbers   63

Chapter 5   Calculus   71

Chapter 6   Differences   87

Chapter 7   Calculator Work   103

INTRODUCTION

This is a second collection of ideas used by a group of teachers in their A level courses. Much of the material has also been used for initial teacher training and for in-service courses.

The members of the group have been attempting to use some investigative and problem solving approaches in their sixth form teaching. They have been looking for ideas directly involved with the present content of the A level syllabuses, or very closely connected to it. They have worked on the principle that the A level syllabus does not appear to be changing fundamentally to allow for mathematical processes to be assessed, but they would like these processes to be part of the learning of their students.

The group has met in Manchester since May 1983 to swap ideas and to suggest new ones. They have now reached a point where it seemed appropriate to offer this new collection to a wider audience.

The group has tried to respond to the challenge presented by paragraph 243 in the Cockcroft Report and to comments in the sixth form chapter about the adult ways of working etc. We have also tried hard not to forget the comment "Mathematics lessons in secondary schools are very often not about anything." (paragraph 462).

Members of the group have welcomed the opportunity to be able to meet and discuss classroom strategies for A level work, the most neglected area of the curriculum. Teachers spend a great deal of time discussing work for low attainers, work with reluctant 5th years etc., but rarely talk about A level. Although sixth form groups are taught by qualified mathematicians HMI comments that A level lessons observed are generally the most arid that they see, and 'teacher telling' plays a predominant part. We accept that this is often because teachers feel under pressure from the syllabus and the need to help students achieve high grades. For this reason most of our investigations are concerned with core areas of the syllabus, although in this second volume we have included some ideas which we believe to be valuable excercises for pupils even though they do not, as such, appear on the syllabus. Students are being encouraged to find things out for themselves and we believe that working in this way helps them to develop confidence and consequently performance.

We hope this collection of starters will be useful to you as an individual teacher, and also be the basis of discussion with colleagues. We have deliberately put these together the way we used them without spending much time 'polishing' them up. In some cases teaching notes and/or lesson reports are included immediately after the worksheets to which they refer, other starters have been left for you to use as you see fit. Where appropriate we have included some examples of pupils' work.

We therefore invite you to use them, and to amend them as you wish. The pupil worksheets may be photocopied for use in your institution.

Chapter 1   **Graphs and Functions**

    1.1   Transformation and Graphs 1

    1.2   Transformation and Graphs 2

    1.3   Transformation and Graphs 3

    1.4   Transformation and Graphs 4

    1.5   Transformation and Graphs 5

    1.6   Transformation and Graphs 6

    1.7   Transformation and Graphs 7

    1.8   Transformation and Graphs 8

    1.9   Transformation and Graphs 9

    1.10   Inverse Functions

    1.11   The Modulus Function

    1.12   Graph Puzzle Sheet 1

    1.13   Graph Puzzle Sheet 2

    1.14   Graph Puzzle Sheet 3

    1.15   Graphs of Rational Functions

    1.16   Sketching Functions 1

    1.17   Sketching Functions 2

    1.18   Sketching Functions 3

    1.19   Funny Functions 1

    1.20   Funny Functions 2

    1.21   Funny Functions 3

    1.22   Combining Modulus Functions

    1.23   Modulus Function Puzzle Sheet

    1.24   Fixing Functions

    1.25   Transforming Graphs

    1.26   Scaling Means and Variances 1

    1.27   Scaling Means and Variances 2

Graphs and Functions

Introduction

This chapter is long and is designed so that sections can be done at intervals throughout a two year A level course. In general, sheets 1 - 14 are suitable for the first year and 15 - 27 are more suited to second year students. The headings of the sheets indicate suitable subsections. Many of the sheets can also be taken by themselves to form part of a completely different sequence of work.

In the past much of the work included in this chapter has been inaccessible to students because plotting and even sketching graphs is so time consuming that it is unlikely that students would have had the patience to sketch enough graphs to get the feel of the general pattern that emerges. Now that computers with good graph plotting programs are readily available ideas of transformations of graphs and investigations of different functions are within the grasp of A level students. We would, therefore, strongly advise the use of a computer for many of these activities even where it is not explicitly mentioned.

Any graph plotting program comes with full instructions for its use and the few explicit references to computer instructions can easily be adapted to different programs. The program referred to here is **'F.G.P.'** which is on the A.T.M. disc, 'Some Lessons in Mathematics With a Microcomputer' (SLIMWAM).

Transformations and Graphs: Descriptions of Worksheets

Sheets 1 - 6 introduce many of the ideas and techniques that are exploited in the rest of the chapter.

1 to 4 require students to investigate transformations of the graphs of $f(x) = x^2$ and $f(x) = \sin x$ using a combination of plotting and sketching. We feel that this experience is important before using the computer.

5 - 6 ask the students to attempt to generalise to other functions using the computer. Without the computer they would not be able to investigate enough functions to convince themselves of the generality.

7 - 11 are a selction of activities to do with transformations of the graph of $f(x) = x^2$ and other graphs. Again a computer is very valuable for these though the students could get some way through them using sketch graphs. A brief lesson report and a student's work is included here.

12 - 14 are called puzzle sheets. The purpose of these is to consolidate the ideas covered. The students are asked to fit equations to graphs. No scales are given so they will obviously get different equations according to how they label the axes. When they have reached conclusions themselves it is valuable to ask them to check each other's equations so that they can appreciate the variety of different possibilities. A computer can also be used in checking.

15 begins to involve more complex ideas in the second part. A computer is valuable.

16 - 18 involve sketching functions without knowing their equations. The first activity revises the transformations discovered in the first few sheets and then goes on to investigate some stranger transformations.

19 - 24 look at some unfamiliar and different types of functions. They begin to look at harder concepts, e.g. continuity. A lesson report is included here.

25 - 27 look at transformations that link directly to statistics work.

1.1 Transformation and Graphs 1

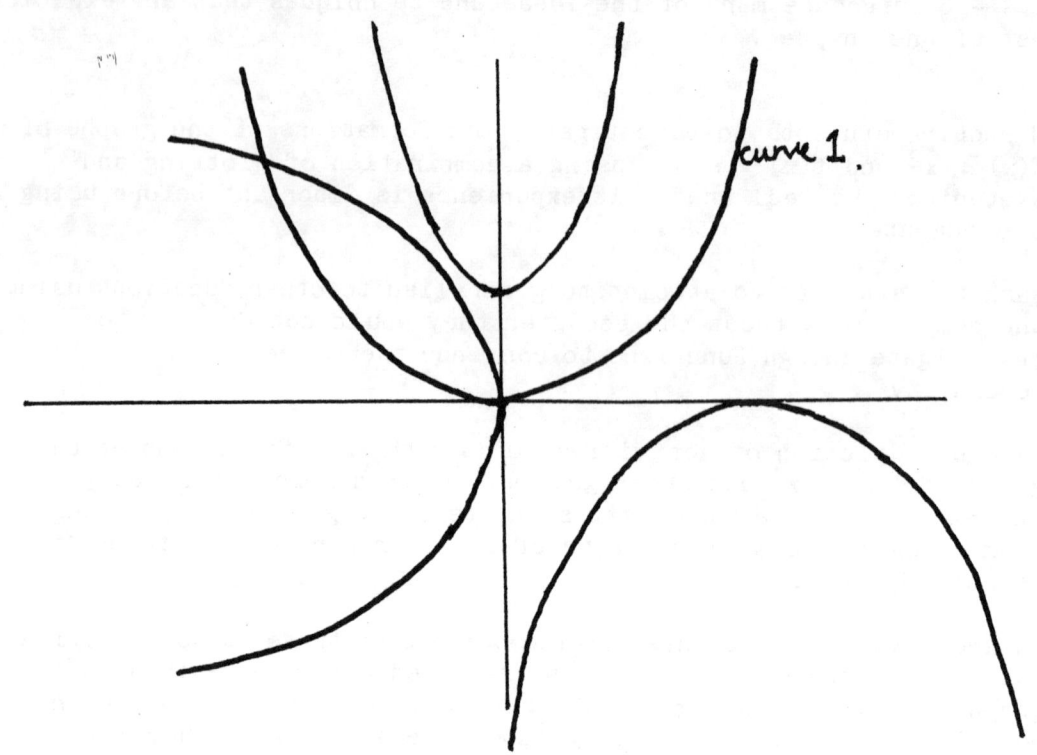

This diagram shows a set of curves that have characteristics in common.

- Describe what is 'the same' about the curves.
- Look at curve 1 and describe in words how you would have to transform it in order to superimpose it on each of the other three curves in turn.

Curve 1 might be the graph of $f(x) = x^2$.

In order to fit equations to correspond to these transformations you will need to investigate the effect of changing $f(x) = x^2$ in various ways ..... (sheet 2)

Transformation and Graphs 2   1.2

INVESTIGATE $x \to af(x)$

On a piece of graph paper draw axes to show $-8 \leq x \leq 8$ and $-60 \leq y \leq 60$.

Consider the function $f(x) = x^2$ and plot the graph of $y = f(x)$.

On the same axes draw the graph of $y = 2x^2$. This can be called the graph of $y = 2f(x)$.

- What transformation do you have to do to the first graph to obtain the second?

Try some other graphs in the form $y = ax^2$, where a has numerical values. (Don't forget that a can be positive, negative or fractional.)

- Describe the transformation for any a.

11

1.3 Transformation and Graphs 3

INVESTIGATE the effect of the following transformations of the graph of $f(x) = x^2$ using SKETCH GRAPHS.

$x \to f(ax)$

    Sketch the graphs of $y = f(2x)$, $y = f(\frac{x}{2})$, $y = f(-3x)$.
    Describe the transformations.
    Generalise to $y = f(ax)$.

$x \to f(x+a)$

    Try $x \to f(x-1)$, $x \to f(x-2)$, $x \to f(x+2)$
    Generalise to $x \to f(x+a)$.
    Describe the transformation.

$x \to f(x) + a$

    Try positive and negative values of a.
    Generalise.
    Describe the transformation.

Transformation and Graphs 4   1.4

INVESTIGATE the effect of the same transformations as on the previous sheet on the graph of $f(x) = \sin x°$

$$x \to af(x)$$

$$x \to f(ax)$$

$$x \to f(x+a)$$

$$x \to f(x)+a$$

On your graph paper use axes to show $-360 \leq x \leq 360$ and $-6 \leq y \leq 6$

- Do the same things happen under the transformations as they did when $f(x) = x^2$?

- Try and reconcile any differences.

1.5 Transformation and Graphs 5

USE A COMPUTER and a graph plotting program to investigate whether these transformations are generally valid for graphs of different $f(x)$.

These instructions are for FGP.

INVESTIGATE $x \to af(x)$.

Let $f(x) = (x-3)(x+2)$.
Enter $y = (x-3)(x+2)$ and press q for 'quick plot'.
Enter $y = 2(x-3)(x+2)$ and press s to obtain this graph superimposed on the other.

- Is the transformation the same as it was when $f(x) = x^2$?
  Superimpose the graphs of $y = -3f(x)$, $y = 0.5f(x)$ and a few more.
- Important points on a graph are turning points and intersections with the axes. What happens to these?
- Describe the transformation $x \to af(x)$ in words.

REPEAT this for different $f(x)$.
Try $f(x) = \tan x$
$f(x) = x^2(x-3)$.

- Is the transformation still the same?

Transformation and Graphs 6 1.6

A. INVESTIGATE $x \to f(ax)$ using a computer.

Enter $y = (x-3)(x-1)(x+2)$.
Press p to plot the graph.
Select x from -2 to 4 and y from -5 to 5.
Enter $f(2x)$, $f(3x)$, $f(0.5x)$ in turn and superimpose the graphs by pressing s.

- What has happened to the original graph in each case? (Think about the turning points and intersections with the axes.)

Use some of the functions suggested below, or others of your own choice, to help you describe the transformation which takes the graph of $y = f(x)$ to the graph of $y = f(ax)$

$f(x) = x(x-3)$	x range: $(-5, 5)$, y range $(-10, 10)$
$f(x) = x - 2$	$(-5, 5)$, $(-10, 10)$
$f(x) = \sin x$	$(-6, 6)$, $(-2, 2)$
$f(x) = \frac{x}{x^2 - 1}$	$(-5, 5)$, $(-10, 10)$
$f(x) = \frac{x}{x^2 + 1}$	$(-5, 5)$, $(-1, 1)$

B. INVESTIGATE $x \to f(x+a)$ and $x \to f(x) + a$ in a similar way.

Identify the transformations involved in each case.

1.7 Transformation and Graphs 7

Enter $y = x^2$ and obtain a sketch by pressing q for "quick plot".

- What value of $f(x)$ do you type in to move the graph up 3 units? Type it in and superimpose using s. (You may find the "double replot" facility useful.)

- Move the original graph down 4 units,
  3 units to the right,
  4 units to the left.

- How do you stretch the graph parallel to the y axis? Try some.
  Stretch it parallel to the x axis.

- How do you enlarge the graph by a scale factor of 2, centre the origin?
  What about a scale factor of k?

- Try to recreate the sketch on sheet 1 on the computer.

A lesson on Enlarging Graphs! - Using 1.7 Transformations and Graphs 7

We had done a lot of work on transformations of graphs, translations, sketches, reflections but had never explicitly thought about an enlargement.

I reminded them about what was meant by 'an enlargement centred on the origin' and aksed them to find what equation would produce an enlarged graph.

A few quickly guessed $y = af(x/a)$ and confirmed this using a computer graph plotter using $y = \sin(x)$. Others, less confident tried different versions before they were successful.

I have rather neglected polynomials when doing work on transformations. This became obvious when the students tried to deal with

$$f(x) = x^3 - 2x^2 + 3$$

They were not immediately happy to replace x by $x/2$ and multiply the entire resulting expression by 2!

Jane, like everyone else, did successfuly produce some acceptable rules and a set of graphs.

Enlargements with a centre not on the origin were left as an extension for only a couple of students.

Enlarging Graphs.   Jane

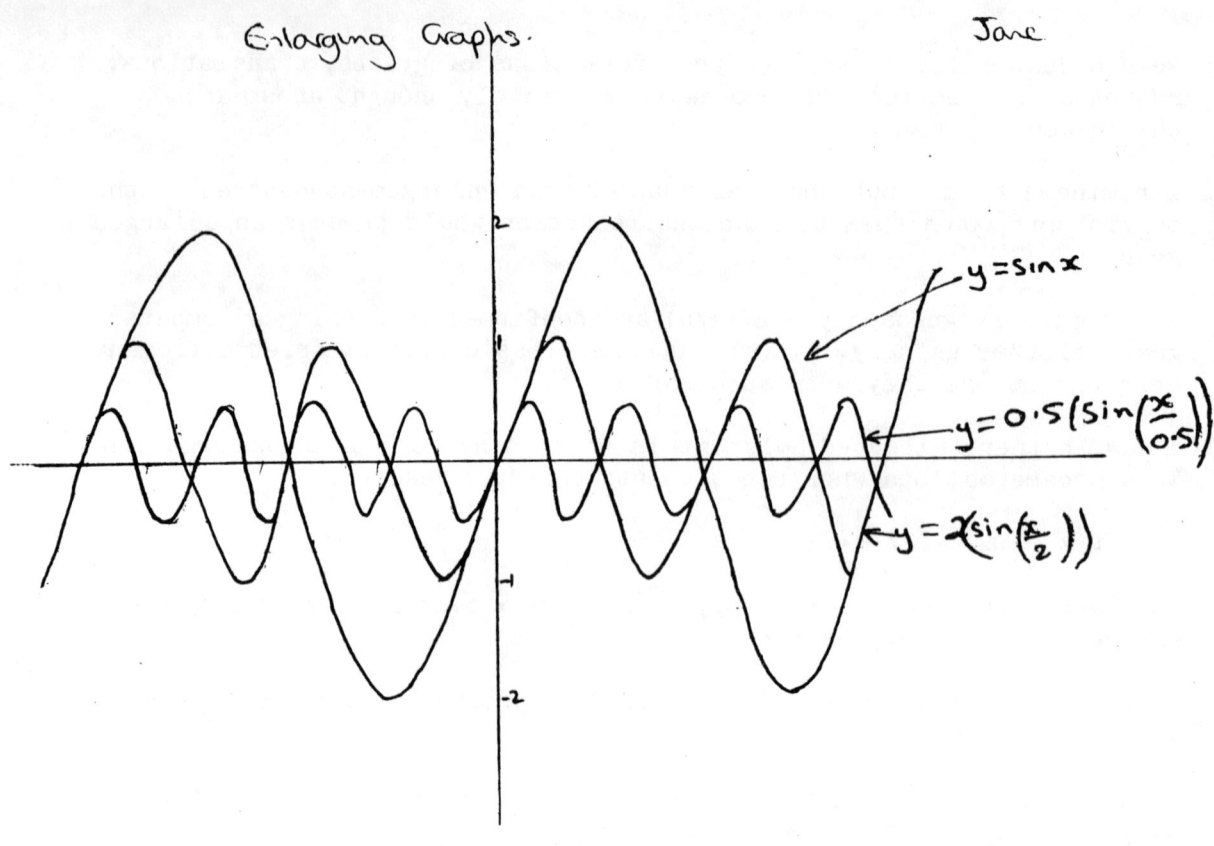

As $y = f(x/a)$ to stretch $\parallel$ to x-axis by a factor of a, and $y = a f(x)$ to stretch $\parallel$ to y-axis by a factor of a

$$y = a(f(\tfrac{x}{a}))$$ in order to enlarge $y = f(x)$ in all directions. This formula can also be used to reduce $y = f(x)$ in all directions by a factor of a.

$y = x^3 - 2x^2 + 3$.

To enlarge $f(x)$, use $y = a(f(\tfrac{x}{a}))$

∴ To enlarge by factor of 2

$y = 2(\tfrac{x}{2})^3 - 4(\tfrac{x}{2})^2 + 6$

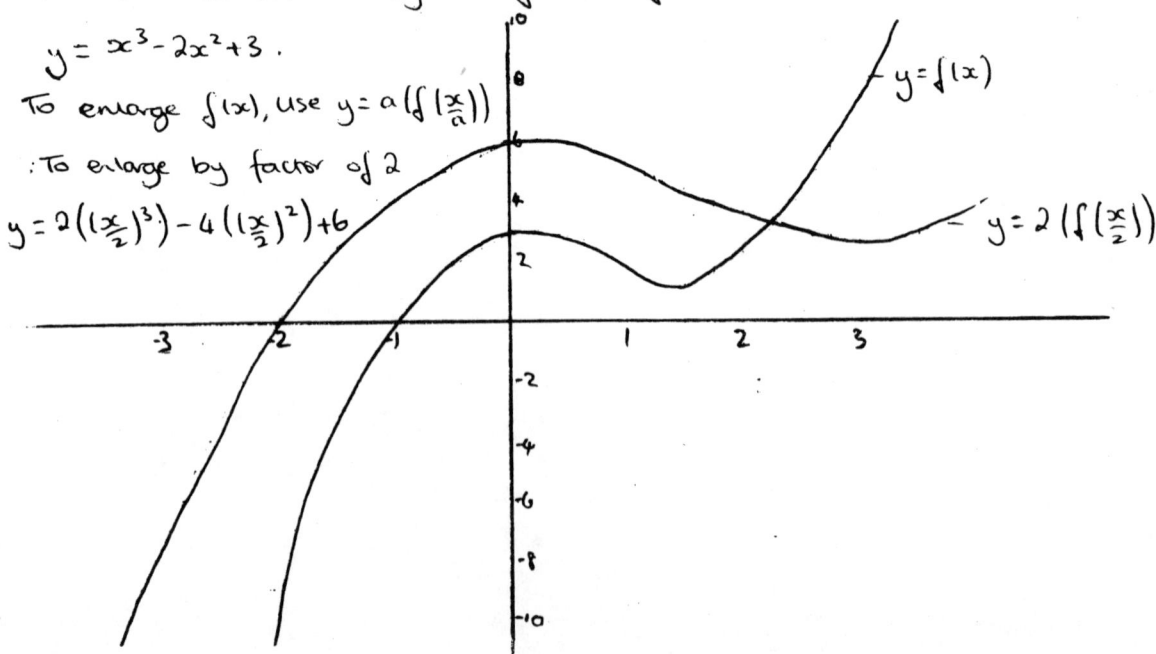

Transformation and Graphs 8   1.8

- What would be the equation of the function whose graph is obtained from that of $y = x^2$ by a translation of
  (i) $\begin{pmatrix} 3 \\ -4 \end{pmatrix}$ (ii) $\begin{pmatrix} \frac{1}{2} \\ -1 \end{pmatrix}$?

- What transformation of the graph of $y = x^2$ would give the graph of $y = x^2 + x - 6$?

- What about $y = 2x^2 - 8x + 4$ from $y = x^2$?

- Try some other quadratics starting with $y = x^2$.

Transform the graph of $y=x^2$ so that it goes through the points $(0,2)$, $(2,2)$ and $(1,-1)$.
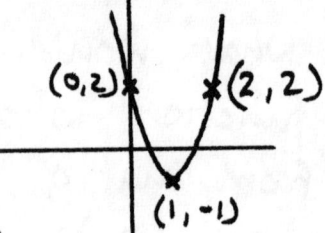
- What transformations did you use?
- What is the equation of the graph?

Find a quadratic graph which goes through $(-1,0)$, $(3,0)$ and $(1,6)$.

Make up some others for yourself.

- Can you outline your general strategy for solving this problem?

Plot the graph of $y = \sin x$ $\quad 0 \leq x \leq 360$.

Find a quadratic graph which fits the first part of the curve $(0 \leq x \leq 180)$.
- Is it a good fit?
- Can you improve it?

You may like to go on to fit a cubic curve to both bends of the sine graph you have plotted.
- How close can you get?

Inverse Functions   1.10

Plot the graph of a function, using the same scales on each axis.

You might try, for example

$$f(x) = 3x - 2$$
$$f(x) = 10 - x$$
$$f(x) = x^2$$
$$f(x) = \frac{1}{x}$$
$$f(x) = \frac{1}{x-1}$$

Superimpose the graph of the inverse function.

- What do you notice about the shapes of the two graphs?
- Where, if anywhere, do they intersect?
- Where do they cut the axes?

1.11 The Modulus Function

Plot the graph of $y = x^2 - 2x$.
Superimpose the graph of $y = abs(x^2 - 2x)$.
Compare the two graphs.

Plot $y = \sin x$ and superimpose $y = abs(\sin x)$.
Compare the graphs.

Repeat for other functions of x.
What happens when you replace $f(x)$ by $abs[f(x)]$?

Can you find any function for which the graph of $y = f(x)$ is the same as the graph of $y = abs[f(x)]$?

Plot the graph of $y = x^2 - 2x$.
Superimpose the graph of $y = [abs(x)]^2 - 2\,abs(x)$.
How do the two graphs compare?

Plot the graph of $y = \sin x$.
Superimpose the graph of $y = \sin[abs(x)]$.
Compare the graphs.

Choose some graphs of your own and describe what happens when you replace x by $abs(x)$ in their equations.
Can you find any graphs for which it makes no difference?

Graph Puzzle Sheet 1   1.12

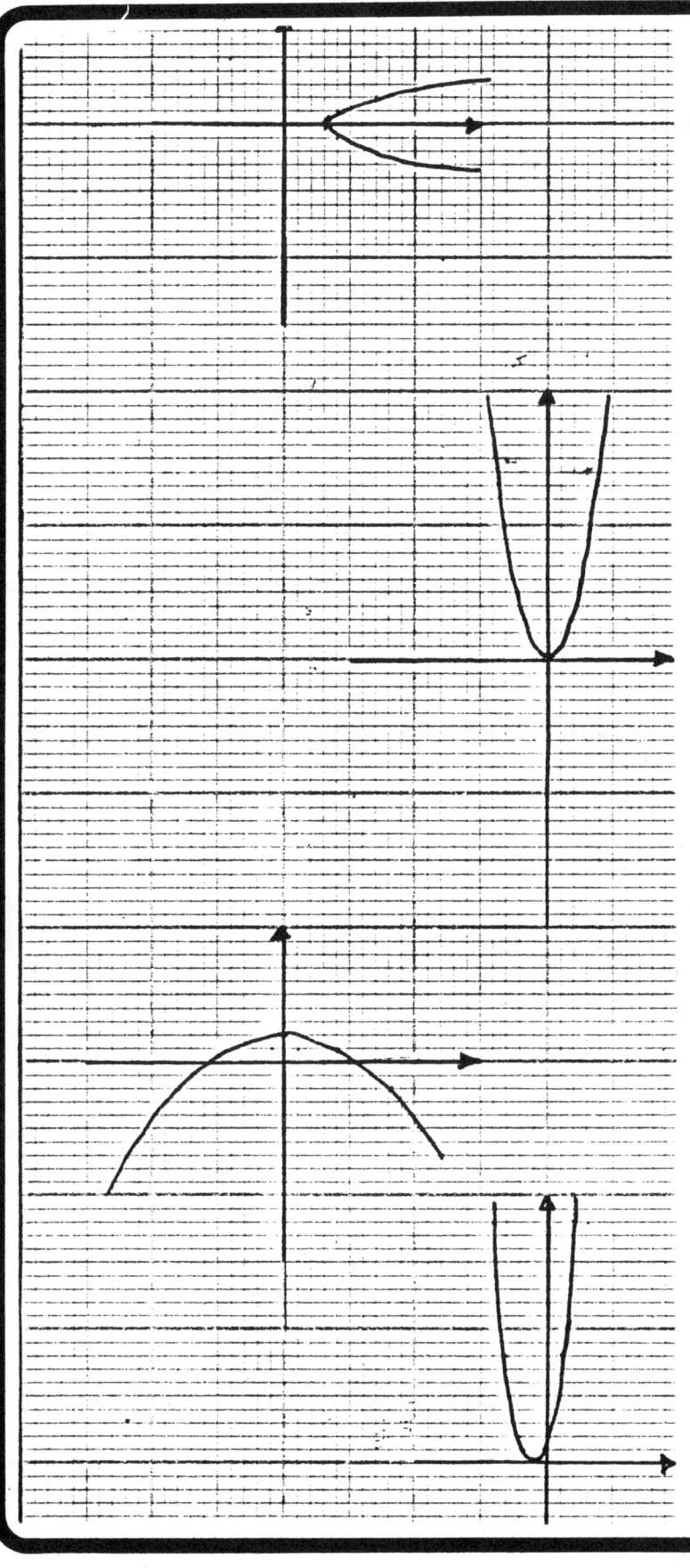

Opposite are the graphs of four quadratics. Their equations are included in the list below:

$f(x) = 8(1-x^2)$
$f(x) = \frac{1}{4}(2-x^2)$
$y^2 = x$
$y = x^2$
$y^2 = x-3$
$f(x) = (x-1)^2$
$f(x) = 2(x+1)^2$

When you have matched the equations to these graphs, try to fit equations to the graphs on puzzle sheets 2 and 3.

1.13 Graph Puzzle Sheet 2

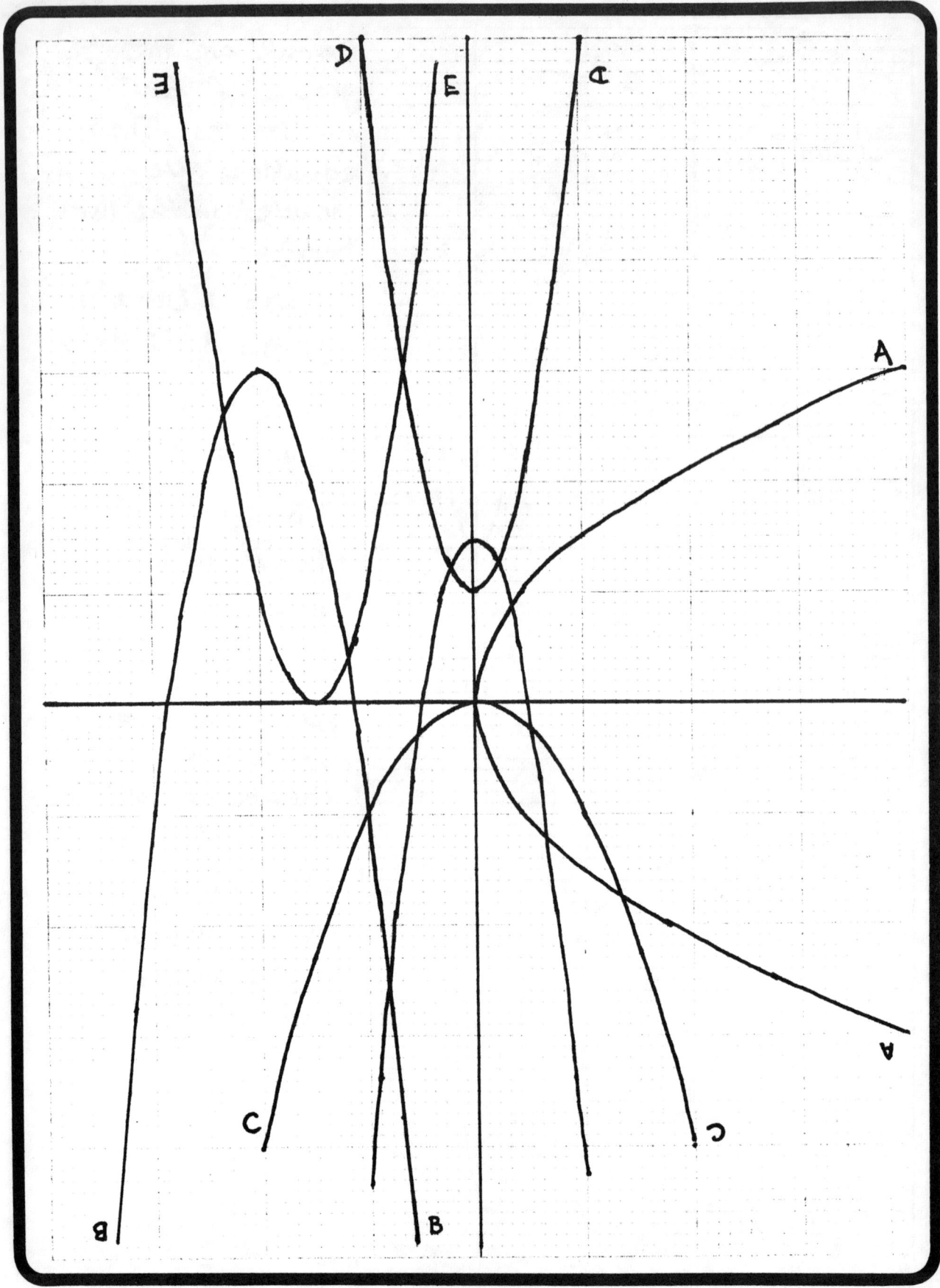

Use either way up.

Graph Puzzle Sheet 3

1.14

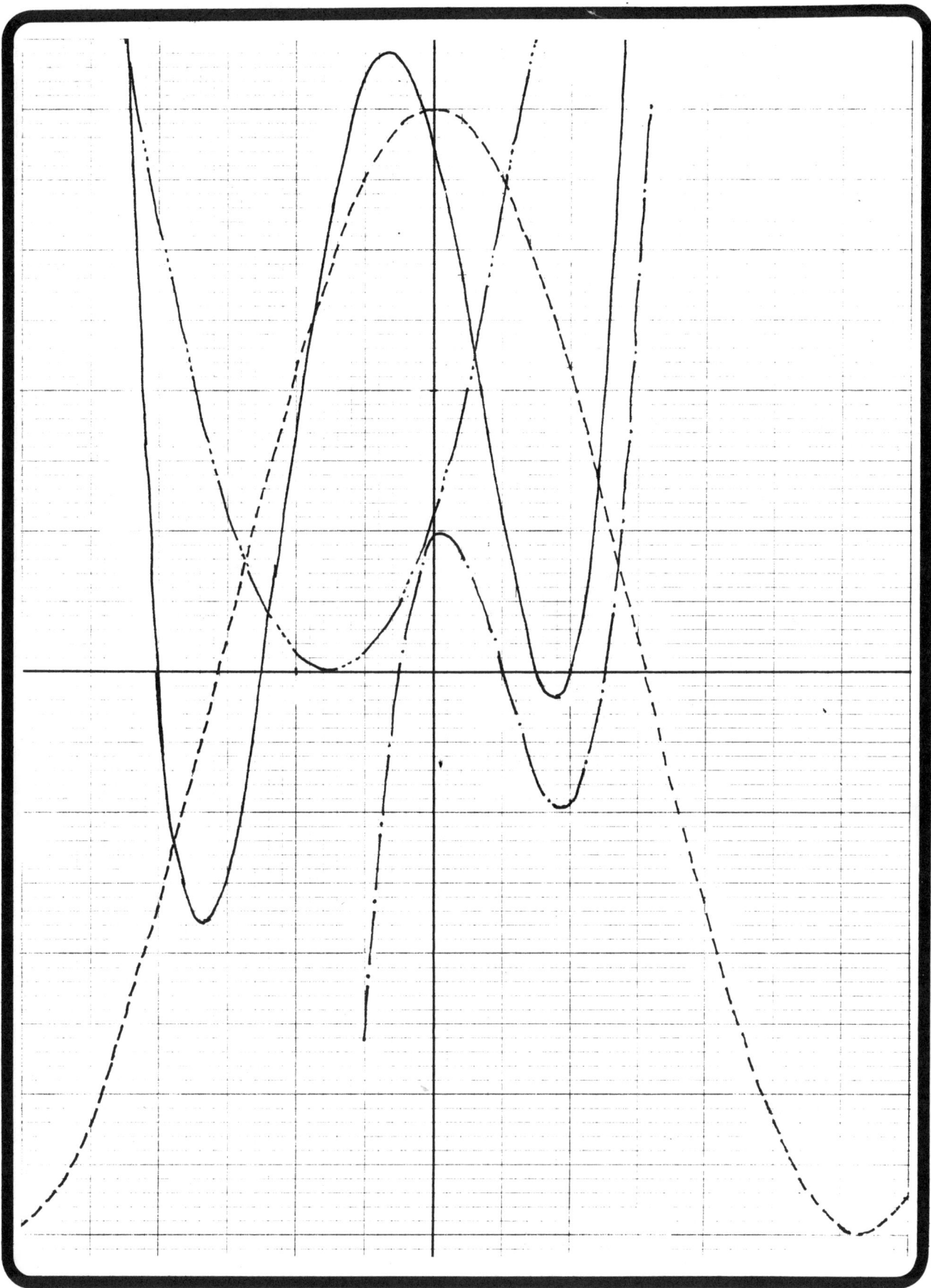

25

1.15 Graphs of Rational Functions

Use a computer and graph plotting program.

A. Plot the graph of any function of the form
$$f(x) = \frac{ax+b}{cx+d}.$$
Start with something simple, maybe letting $a=0$.
e.g. $f(x) = \frac{1}{x+2}$.

Investigate the effect of varying the values of a, b, c and d.

Find out what you can about graphs of functions of this form.

B. Now try functions of the form
$$f(x) = \frac{ax^2+bx+c}{dx^2+ex+f}.$$
You may prefer to start with the numerator and denominator in factorised form.

Investigate the graphs produced for different values of a, b, c, d, e, f.

Each time make a note of the important features of the graph.

Sketching Functions 1   1.16

A
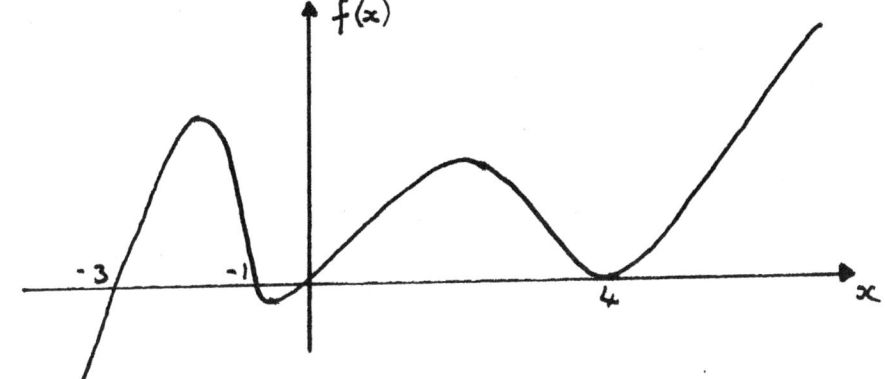

Here is a function $f(x)$ described only by its graph. Describe the transformations which will give the graph of each of the following. Sketch each graph.

(1) $f(x)+3$
(2) $f(x+3)$
(3) $f(x)-1$
(4) $f(x-1)$
(5) $f(3x)$
(6) $f(\tfrac{1}{2}x)$
(7) $f(-x)$
(8) $-f(x)$
(9) $-f(-x)$
(10) $f(\text{MOD}(x))$
(11) $f(-\text{MOD}(x))$
(12) $\text{MOD}[f(x)]$
(13) $-\text{MOD}[f(x)]$
(14) $-\text{MOD}[-f(x)]$

B. Sketch the graph of $y = \log_e(x)$. Apply all the above transformations to this graph.

C. For which of the transformations in section A can a graph remain exactly as it was before? For each of these transformations sketch a graph which would stay the same.

1.17 Sketching Functions 2

Sketch the graph of $f(x) = x^2 - 9$.
Draw in the lines $f(x) = 1$ and $f(x) = -1$.
We are going to build up the graph of $g(x) = \dfrac{1}{x^2 - 9}$.
Sketch the graph of $g(x)$ in a different colour.

(1) For what values of x is $g(x)$ undefined?
Draw in these asymptotes.

(2) What is the value of $g(x)$ if $f(x) = 1$?
Plot the graph of $g(x)$ at such points?

(3) What is the value of $g(x)$ if $f(x) = -1$?
Plot the graph of $g(x)$ at such points.

(4) If $f(x)$ is increasing what is happening to $g(x)$?
If $f(x)$ takes a minimum or maximum value what happens to $g(x)$?

(5) If $f(x) > 1$ what range of values is possible for $g(x)$. Use this and question 4 to sketch $g(x)$ for all values of x such that $f(x) > 1$.

(6) Repeat question 5 for $f(x) < -1$, $0 < f(x) < 1$, $0 > f(x) > -1$.

Your sketch of $g(x)$ should now be complete.

Sketching Functions 3   1.18

The graph of h(x) is drawn below.

By working through the same process as on the previous sheet, sketch the graph of $\dfrac{1}{h(x)}$.

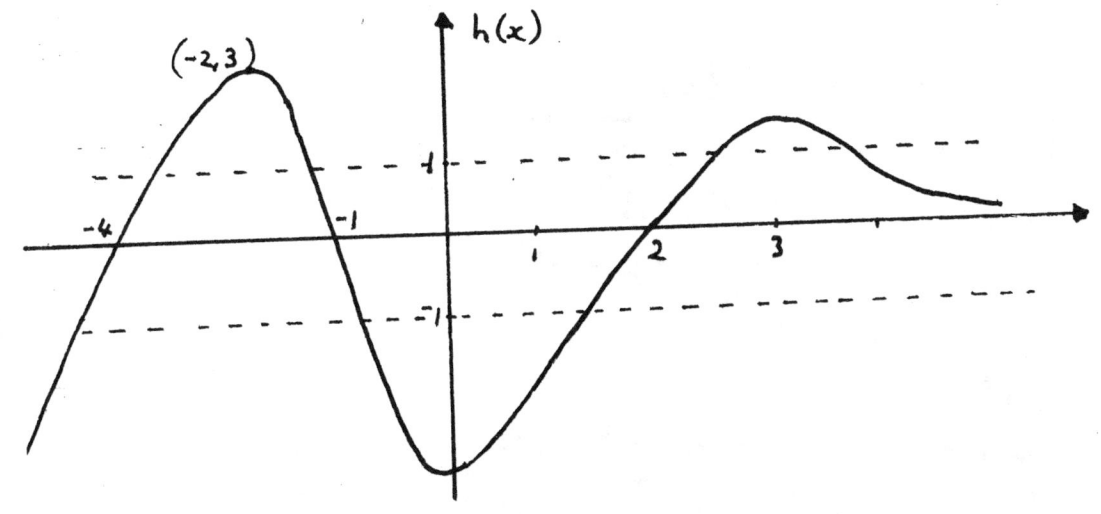

Funny Functions 1

The first funny function we look at is $f(x) = INT(x)$. $INT(x)$ stands for the largest integer less than or equal to x.

So $INT(2) = 2$, $INT(2.9) = 2$, $INT(-2.9) = -3$.

Sketch the graph of $f(x) = INT(x)$.

Now sketch the graphs of the following functions involving $INT(x)$.

$$f_1(x) = x - INT(x)$$
$$f_2(x) = \sqrt{x - INT(x)}$$
$$f_3(x) = INT(x) + \sqrt{x - INT(x)}$$
$$f_4(x) = (x - INT(x))^2$$
$$f_5(x) = INT(x) + (x - INT(x))^2$$
$$f_6(x) = INT(x^2)$$
$$f_7(x) = INT(2x)$$
$$f_8(x) = INT(x) + INT(-x)$$
$$f_9(x) = INT(x) + \sqrt{x - INT(x)} \quad ; \; INT(x) \text{ even}$$
$$ INT(x) + (x - INT(x))^2 \quad ; \; INT(x) \text{ odd}$$
$$f_{10}(x) = INT\left(\frac{100}{x}\right)$$

Are any of these functions continuous everywhere?
Are any of these functions continuous at some points?

Funny Functions 2   1.20

A. Can you draw $f_1(x) = \begin{cases} x & x \text{ is rational} \\ x & x \text{ is irrational} \end{cases}$?

Is it continuous?

Try these similar functions:
$$f_2(x) = \begin{cases} 1+x & x \text{ rational} \\ 1 & x \text{ irrational} \end{cases}$$
$$f_3(x) = \begin{cases} x - \text{INT}(x-\tfrac{1}{2}) & x \text{ rational} \\ 1 & x \text{ irrational} \end{cases}$$
$$f_4(x) = \text{largest factor of } x < |x|, \ x \in \mathbb{Z}$$

Are any of them continuous anywhere?

B. Here are some more oddities to sketch and comment on.

$$g_1(x) = \begin{cases} \cos \tfrac{1}{x} & x \neq 0 \\ 0 & x = 0 \end{cases}$$
$$g_2(x) = \begin{cases} x \cos \tfrac{1}{x} & x \neq 0 \\ 0 & x = 0 \end{cases}$$
$$g_3(x) = \begin{cases} x^2 & x > 0 \\ -x^2 & x \leq 0 \end{cases}$$
$$g_4(x) = x^x \quad \text{(look carefully at what happens between 0 and 1.)}$$

Using **Funny Functions 1 and 2** (1.19 and 1.20)

I started my new further maths group off with this topic. We spent a few minutes discussing what a function is - some of the group were familiar with the word, some not, and <u>none</u> had anything more than a hazy idea of what it meant!

We then worked on "Funny Functions 1" and had many different versions of each sketch on the board, eventually managing to reach a consensus about which one was acceptable.

We then did some work from "Functions, Formulae and Graphs" - Keith Hirst's article in Sixth Form Mathematics. This involved using the modulus function to find a single description for:

$$f_1(x) = \begin{cases} x \text{ if } x \geqslant 0 \\ 0 \text{ if } x < 0 \end{cases} \qquad f_2(x) = \begin{cases} 0 \text{ if } x \geqslant 0 \\ x \text{ if } x < 0 \end{cases}$$

$$f_3(x) = \begin{cases} 2x \text{ if } x \geqslant 0 \\ 3x \text{ if } x < 0 \end{cases} \qquad f_4(x) = \begin{cases} x^2 \text{ if } x \geqslant 0 \\ -x^2 \text{ if } x < 0 \end{cases}$$

(there is lots more in the article which we have not used yet).

We then went on to look at Funny Functions 2, and will come back to some of the nasties on there now we have got a copy of F.G.P.

They also did a more routine homework on finding values of functions etc. out of Turner Book 1, page 32.

Funny Functions 3

1.21

A. Differentiate each of the following.

$$f_1(x) = \log(\sin x)$$

$$f_2(x) = \log\log(\sin x)$$

$$f_3(x) = \cosh^{-1}(\cos x)$$

B. Sketch the graph of each function.

Comment on the results.

What do your answers to section A mean?

1.22 Combining Modulus Functions

A. Sketch the graphs of
$$y = |x| + |x-1|$$
$$y = |x| - |x-1|$$

Investigate other graphs of the form
$$y = |ax+b| \pm |cx+d|$$

B. Sketch the graphs of
$$y = |x| + 2|x-1|$$
$$y = |x| - 2|x-1|$$

Generalise to other functions of the form
$$y = a|bx+c| \pm d|ex+f|$$

C. Explore combinations of 3 or more modulus functions.
   e.g. $y = |x| + |x+1| + |x-1|$

D. Sketch
$$y = x^2 + |x^2-1|$$
$$y = x^2 - |x^2-1|$$
$$y = |x^2-1| \pm |x^2-4|$$

Look for generalisations.

Modulus Function Puzzle Sheet 1.23

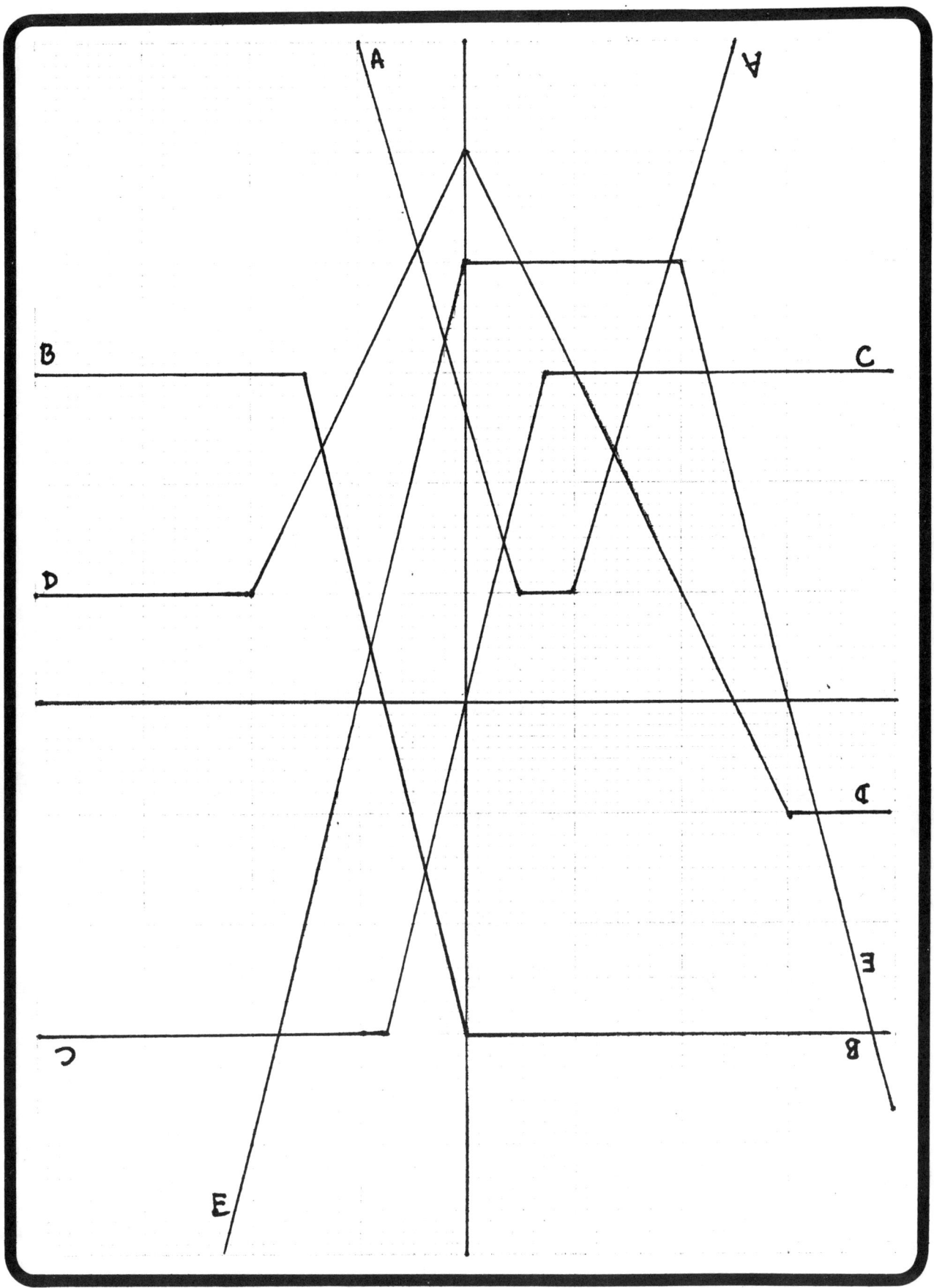

Use either way up.

1.24 Fixing Functions

Answer these — compete with your neighbour.

- Find a function f such that
$$f(1)=0 \quad f(2)=1 \quad f(13) \text{ is as large as possible.}$$

- Find a function g such that
$$g(1)=0 \quad g(2)=1 \quad g(3)=2$$
$$g(13) \text{ is as large as possible.}$$

- Find a function h such that
$$h(1)=0 \quad h(2)=1 \quad h(10)=2$$
$$h(13) \text{ is as large as possible.}$$

Now repeat them all but make the values at 13 as <u>small</u> as possible.

Transforming Graphs 1.25

A. This is the graph of $y = f(x)$.

A transformation is applied to the graph.

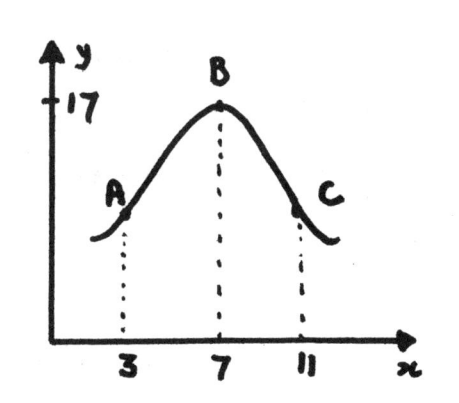

[Graph showing curve with points A at x=3, B at x=7 (y=17), C at x=11]

The resulting graph is:

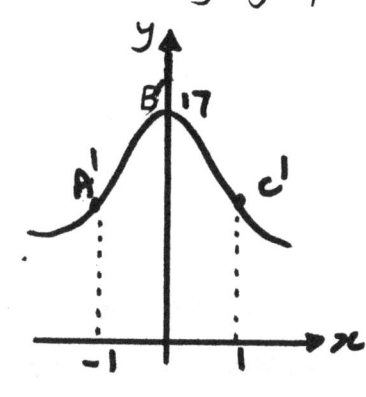

[Graph showing curve with points A' at x=-1, B' at y=17 on y-axis, C' at x=1]

If $A \to A'$, $B \to B'$, $C \to C'$, what function must be plotted to get this graph?

Generalise...

B. Repeat the above for this pair of graphs.

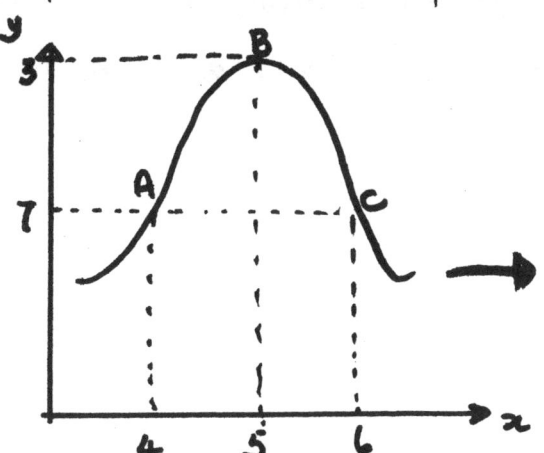

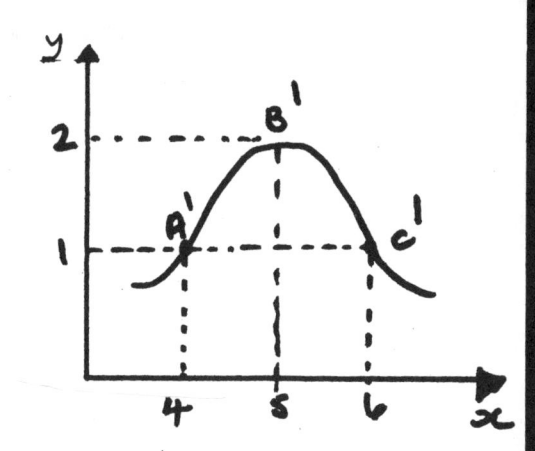

[Left graph: curve with A at x=4 (y=7), B at x=5 (y=3), C at x=6 (y=7). Right graph: curve with A' at x=4 (y=1), B' at x=5 (y=2), C' at x=6 (y=1)]

C. Can you combine both kinds of transformation?

Scaling Means and Variances 1

Find the mean and variance of these numbers.

$$1, 3, 3, 4, 6, 6, 7, 8, 8, 9$$

What happens to the mean and variance if a transformation of the form
$$x \rightarrow ax+b$$
is applied to the numbers.

Try these for some ideas.

The numbers have 4 added to them.
have 4 subtracted from them.
are doubled.
are halved.
are halved and have 4 added to them.
have 4 added to them and are halved.

Scaling Means and Variances 2

A. Find the mean, variance and standard deviation of these numbers.

$$-5, -3, -3, -1, 0, 0, 1, 2, 2, 3.$$

Find other sets of numbers whose mean, variance and standard deviation you can find using these results.

B. Suppose the numbers in A were the midpoints for grouped data?

Comments on **Scaling Means and Variances**

Scaling Means and Variances (1)

The transformation $x \to ax + b$

Some pupils may find it easier to find a set of rules for the cases when either a = 0 or b = 0 before progressing to the more general case.

The ideas on this worksheet link up with many in the section and, in particular, with Scaling Graphs, which is a graphical approach to the same problem.

It is useful to ask the pupils to explain why both a and b affect the mean but only a affects the variance.

Scaling Means and Variances (2)

This worksheet uses the results of the first sheet to show how a change of variable for data may be useful to give an easier set of data to work with.

The worksheet uses this idea in reverse. Here the easier set of data is given and students have to find some sets of data which could have been the starting point.

Chapter 2   **Trigonometry**

    2.1   Trig Formulae 1

    2.2   Trig Formulae 2

    2.3   Trig Function Puzzle Sheet

    2.4   Reciprocals of Trig Functions

    2.5   Inverse Trig Functions

    2.6   Problems With A Bookcase

2.1 Trig Formulae 1

Plot the graph of $y = \sin x \cos x$
on the screen.

Superimpose the graph of $y = \sin x$ and compare the phases, periods and amplitudes of the two graphs.

Transform the graph of $y = \sin x$ to fit the graph of $y = \sin x \cos x$.

Express what you have done algebraically.

Try fitting the graph of $y = \cos x$ to the graphs of $y = \sin^2 x$ and $y = \cos^2 x$.

Express what you have done algebraically.

Trig Formulae 2 — 2.2

Plot the graph of $y = \sin x + \cos x$ on the screen.

Superimpose the graph of $y = \sin x$ and compare the phases, periods and amplitudes of the two graphs.

Transform the graph of $y = \sin x$ to fit the graph of $y = \sin x + \cos x$

Express what you have done algebraically.

Now try another graph of the form $y = a \sin x + b \cos x$ for simple values of a and b.

Can you find ways of predicting which sine graph will fit?

What about fitting cosine graphs to the graphs of $y = a \sin x + b \cos x$?

Trig Formulae

We started by doing some graphs by hand. I asked the class (first year A level further maths) to sketch the graphs of

$$\sin^2 x \qquad \cos^2 x$$
$$|\sin x| \qquad |\cos x|$$
$$\sin x \cos x \qquad \sin x + \cos x$$

The realisation that $\cos^2 x$ and $\sin^2 x$ could never be negative produced graphs something like $|\cos x|$ and $|\sin x|$ at first. When the students then got on to the modulus sketches they doubted their earlier attempts and went back to check. Calculators made it easy to work out actual co-ordinates and students seemed to feel happier with this method than just sketching.

Next lesson we went to the computers (3 for a class of eight) and used F.G.P. from the SLIMWAM disc.

There were various initial problems and discussions - the surprise when using the "quick plot" facility generally produced straight lines not the familiar curves - and I encouraged the class to work fairly freely at first, but aiming at the work on the two worksheets.

Very soon one student told me that $\dfrac{\sin 2x}{2} =$

and the other standard results came out without problems.

It was more difficult to find a rule for an alternative form for $a \cos x + b \sin x$. The work led to lots of useful discussion (and argument!) and hypotheses were suggested and checked.

Later we did some classwork on the expansions of $\sin(x+\alpha)$ and $\cos(x+\alpha)$ so we were able to discuss, refine and discard, if necessary, the various theories.

Trig Function Puzzle Sheet 2.3

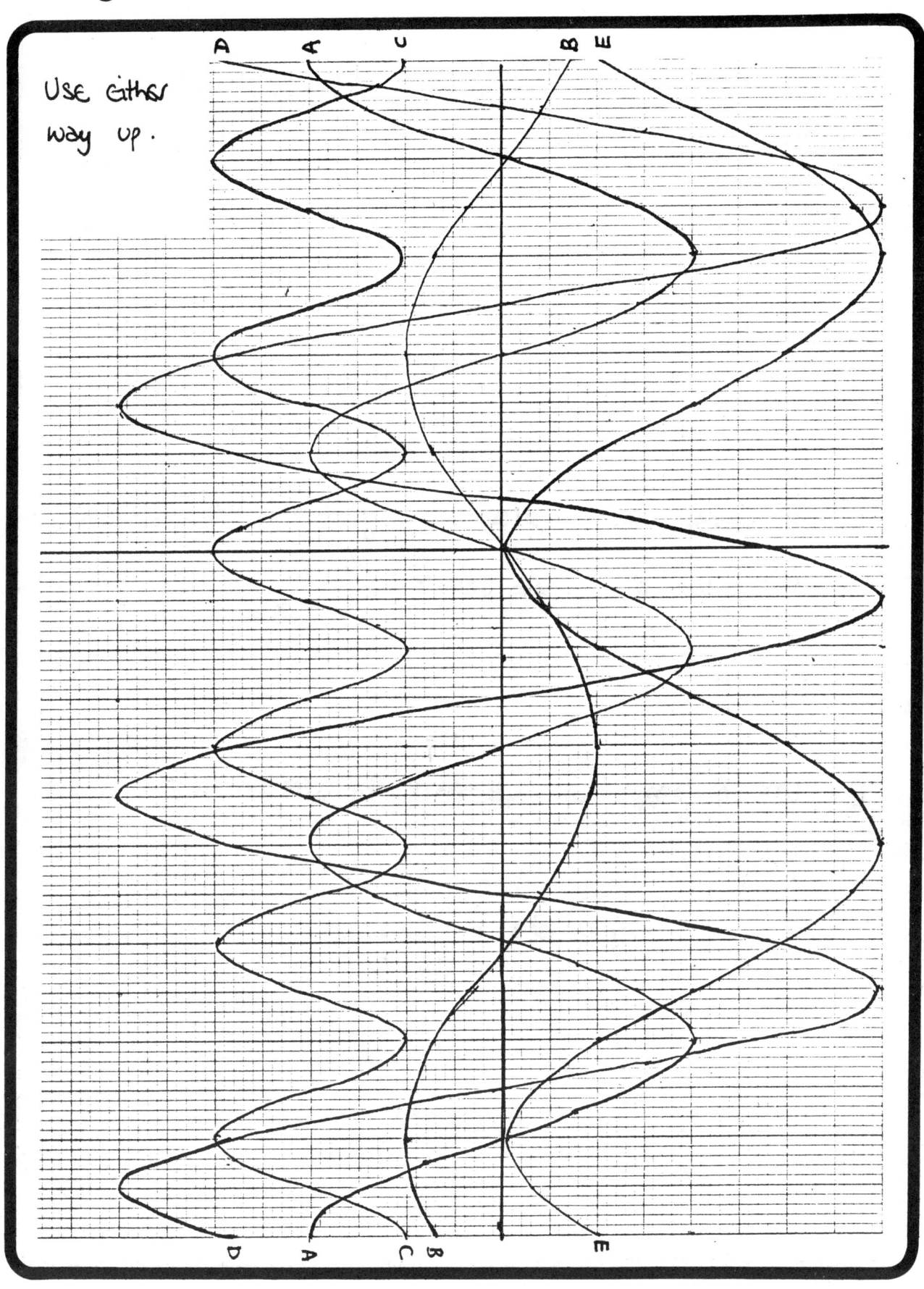

Use either way up.

2.4 Reciprocals of Trig Functions

A   Sketch the graph of $y = \sin x$ for values of x from -2π to 2π.

Use the ideas of Sketching functions 3 to sketch the graph of $y = \text{cosec } x = \dfrac{1}{\sin x}$

B   Similarly, use a graph of $y = \cos x$ to sketch the graph of $y = \sec x = \dfrac{1}{\cos x}$

C   Use a graph of $y = \tan x$ to sketch the graph of

$$y = \cot x = \dfrac{1}{\tan x}$$

Inverse Trig Functions 2.5

Draw the graph of $y = \sin x$ for $-4\pi \leq x \leq 4\pi$. Use the same scale on both axes, despite the problems that this causes.

Use the $\sin^{-1}$ button on your calculator to help you plot the graph of $y = \sin^{-1} x$. Use the same axes but plot in a different colour.

Use the work on transformations of graphs to find the inverse of the function $f(x) = \sin x$.

How is this related to your graph of $y = \sin^{-1} x$? Why do you think the two are different?

Repeat this work using the graphs of $y = \cos x$ and $y = \tan x$.

Inverse Trigonometric Functions

Students should already be familiar with the idea of an inverse function (see Chapter 1) and how it can be drawn using the line y = x.

The worksheet suggests using the same scale for both axes so that the y = x line can easily be seen.

Using the $\sin^{-1}x$ button (or arc sin) on a calculator the principal value of $y = \sin^{-1}x$ can be drawn.

By reflecting y = sinx in the line y = x the entire graph of $y = \sin^{-1}x$ can be found.

Discussion with the pupils should bring out the difference between a function and a mapping, a function being a one-to-one or many-to-one mapping.

The entire graph of $y = \sin^{-1}x$ is a one-to-many mapping but by restricting the range to the principal value we obtain a function.

Problems with a Bookcase

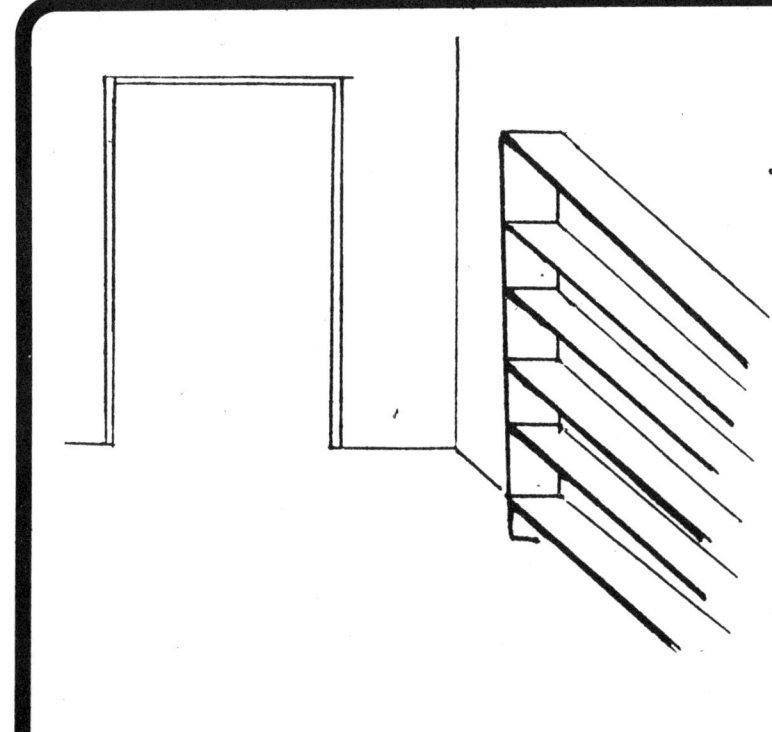

The doorway is 2m high and 0.8m wide.

The bookcase is 2·05m tall and quite long. Will it fit through the door?

Some questions to investigate:

What happens as you tilt the bookcase to try to get it through the door?

What is the locus of a point on the top front edge... the top back edge?

Will these points clear the ceiling if it is 2·1m high?

What is the minimum height of ceiling for which you can tip the bookcase?

Difficulties With A Bookcase

Once a diagram has been produced,

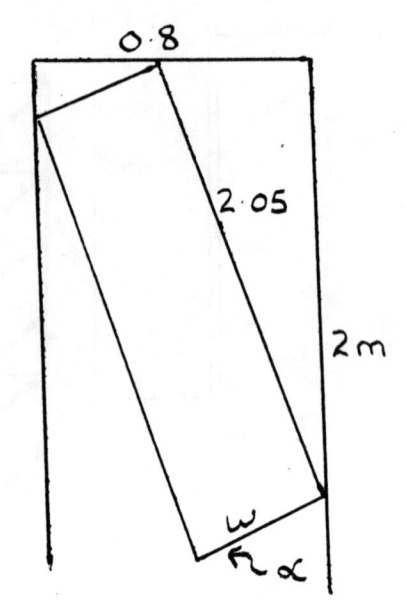

$2 = w \sin\alpha + 2.05\cos\alpha$

$0.8 = w \cos\alpha + 2.05\sin\alpha$

so;

$w = (2 - 2.05\cos\alpha)/\sin\alpha$

and $w = (0.8 - 2.05\sin\alpha)/\cos\alpha$

These can be graphical and the instruction sought, by calculator or computer.

A simple crude computer method is to tabulate each expression over a range of values - compare them, then retabulate over a smaller range.

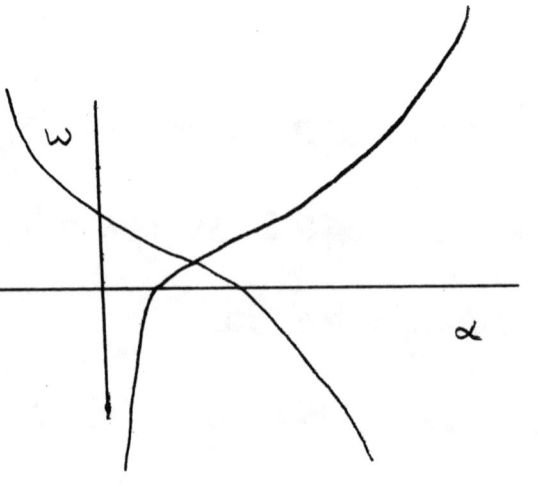

Other numerical methods are also obviously feasible (but are they quicker?)

A continued investigation of the geometrical configuration can lead to three important formulae!

If students are allowed plenty of time to play with the situation themselves and are encouraged to reflect on the situation then, even if these are not, discovered, teacher exposition should then be accessible to the student.

Relative to O, the co-ordinates of
P are given by various equations;

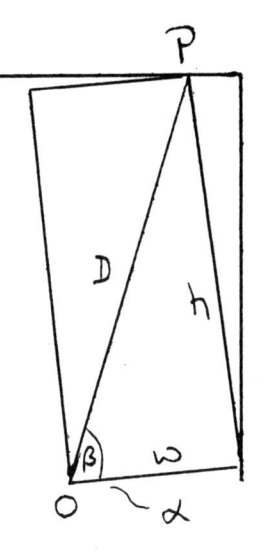

Bookcase hitting low ceiling!

$$y = w\sin\alpha + h\cos\alpha$$
or $$y = D\cos\beta \sin\alpha + D\sin\beta \cos\alpha$$ ✳
or $$y = D\sin(\alpha + \beta)$$

and $$x = w\cos\alpha - h\sin\alpha$$
or $$x = D\cos\beta \cos\alpha - h\sin\beta \sin\alpha$$ ✳
or $$x = D\cos(\alpha + \beta)$$

Also, looking at:

$$y = w\sin\alpha + h\cos\alpha \quad \text{and} \quad y = D\sin(\alpha + \beta)$$

points to the possiblity of manipulating combinations like
$a\sin\alpha + b\cos\alpha$ into $R\sin(\alpha + \beta)$, although some work with a graph plotter
on $y = 2\sin x + 3\cos x$ etc should precede a formal statement.

Chapter 3   Vectors

    3.1   Vectors 1

    3.2   Vectors 2

    3.3   Vectors and Movement

    3.4   Circular Motion

3.1 Vectors 1

Consider two points A and B such that
$$\vec{OA} = \begin{pmatrix} x_1 \\ y_1 \end{pmatrix} \text{ and } \vec{OB} = \begin{pmatrix} x_2 \\ y_2 \end{pmatrix}$$
where O is the origin

find $\vec{OP}$ if

- P is the midpoint of $\vec{AB}$
- P divides $\vec{AB}$ in the ratio $3:4$
- P is beyond B and $BP = \frac{1}{3} AP$
- P divides AB in the ratio $m:n$

find $\vec{OP}$ if P is a variable point anywhere on the line through A and B.

Your last answer is the <u>vector equation</u> of AB
Now write down the vector equation of the line joining $(5,3)$ and $(2,1)$.
Try some other pairs of points.

Repeat this work for 3 dimensions, and thus find the vector equation of the line joining (x_1, y_1, z_1) and (x_2, y_2, z_2).

Vectors 2   3.2

- Under what conditions are the two vectors $\begin{pmatrix} x_1 \\ y_1 \end{pmatrix}$ and $\begin{pmatrix} x_2 \\ y_2 \end{pmatrix}$ perpendicular to each other?

- What is the angle between the two vectors $\begin{pmatrix} x_1 \\ y_1 \end{pmatrix}$ and $\begin{pmatrix} x_2 \\ y_2 \end{pmatrix}$?

- If the first vector is resolved in the direction of the second, what is the size of the component?

- If the second vector is resolved in the direction of the first, what is the size of the component?

- What is the area of the triangle ABC, where $\vec{AB} = \begin{pmatrix} x_1 \\ y_1 \end{pmatrix}$ and $\vec{AC} = \begin{pmatrix} x_2 \\ y_2 \end{pmatrix}$?

- Now repeat all this in three dimensions!

3.3 Vectors and Movement

The position vector of a point P $\underline{r}(t) = \begin{pmatrix} t^2 \\ 4t \end{pmatrix}$, where t represents time in seconds and distances are in metres.

Draw a diagram to illustrate the motion of P over a period of time. Where was P two seconds ago?

What is $\underline{r}(3) - \underline{r}(2)$?

What interpretation can you give to this vector?

What about
$\underline{r}(5) - \underline{r}(4)$
$\underline{r}(4.5) - \underline{r}(4)$
$\underline{r}(4.1) - \underline{r}(4)$

and
$\underline{r}(6) - \underline{r}(5)$
$\underline{r}(5.5) - \underline{r}(5)$
$\underline{r}(5.1) - \underline{r}(5)$
⋮

Comment on the magnitudes and directions of these vectors

When is P moving m quickly, when t = 4 or t = 5? How fast is it moving? In which direction?

What vector can represent this?

Generalise!

What about other values of t?

What about other functions of t?

What about acceleration? and 3 dimensions?

Circular Motion 3.4

A particle P is moving round a circle of radius 6m with an angular velocity of 2 radians per second.

What is the position vector $\underline{r}$ of the particle as a function of time, t, in seconds?

What is P's velocity ..... its acceleration?
What are their directions and magnitudes?

Generalise!

to a circle of radius R metres and angular velocity ω radians per second.

What is the velocity $\underline{v}$?
What is its direction? and magnitude?

What is changing?   What is constant?

What about acceleration $\underline{a}$?
Direction? magnitude?

Vectors 1 and 2

I used these sheets with an A level group in the summer term of their first year. As some of them had not met vectors at all before, and others had forgotten their previous work, I also provided an O level type sheet for them to do first if they wanted to.

They worked in small groups, discussing and arguing until satisfied with their results.

We discussed sheet 1 together afterwards.

Sheet 2 threw up the expression $x_1 x_2 + y_1 y_2$ so many times that it seemed natural to have a name and notation for it. I don't think that scale or product has caused them any trouble since this experience.

Movement and vectors / Circular Motion

Despite the similarity with earlier work on parametric representations many students struggled to produce an acceptable diagram of the path of P. It thus provided both valuable revision and an experience of having a different viewpoint.

Discussion of their work while it was taking place helped to make the link between an average velocity vector and $\underline{r}(4) - \underline{r}(2)$, when suitably scaled. Most students worked steadily and tediously through the sequence of differences $\underline{r}(5) - \underline{r}(4)$, $\underline{r}(4.5) - \underline{r}(4)$, $\underline{r}(4.1) - \underline{r}(4)$, ... until eventually they made the connection between the process and componentwise differentiation! Paul's work shows this very well. Despite having done this once, when Brian tried working with a 3-D space curve his work shows that he felt the need to repeat this process before differentiating directly!

After this the only difficulty experienced with the circular motion sheet was that of initially setting up equations for the position vector. They immediately differentiated, and the usual formula were trivially obvious.

As this was a part of their applied maths course they were asked to use these results to analyse the motion of earth satellites, assuming they followed a circular path, and the acceleration experienced by bodies standing on the surface of a rotating (nearly) spherical earth.

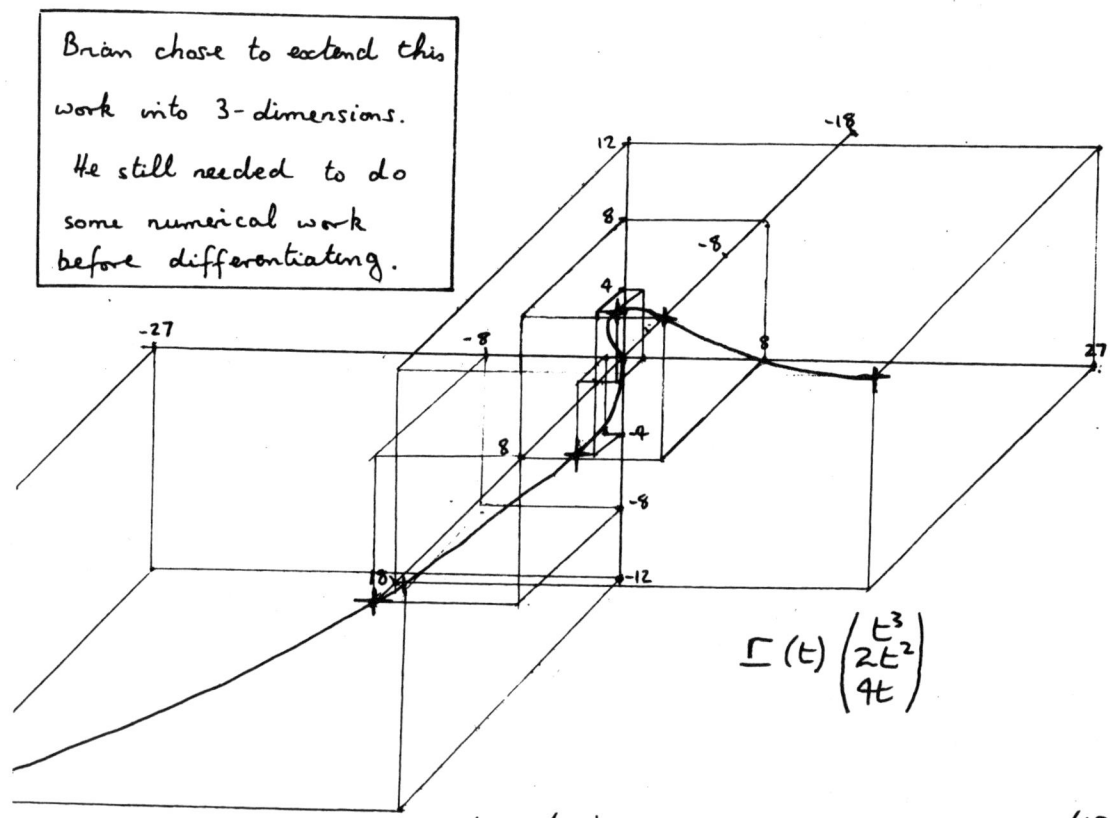

Brian chose to extend this work into 3-dimensions. He still needed to do some numerical work before differentiating.

$$\underline{r}(t) \begin{pmatrix} t^3 \\ 2t^2 \\ 4t \end{pmatrix}$$

i) $\underline{r}(3) - \underline{r}(2) = \begin{pmatrix} 27 \\ 18 \\ 12 \end{pmatrix} - \begin{pmatrix} 8 \\ 8 \\ 8 \end{pmatrix} = \begin{pmatrix} 19 \\ 10 \\ 4 \end{pmatrix}$ ∴ Instantaneous Velocity = $\begin{pmatrix} 19 \\ 10 \\ 4 \end{pmatrix}$

ii) $\underline{r}(2.5) - \underline{r}(2) = \begin{pmatrix} 16.625 \\ 12.5 \\ 10 \end{pmatrix} - \begin{pmatrix} 8 \\ 8 \\ 8 \end{pmatrix} = \begin{pmatrix} 7.625 \\ 4.5 \\ 2 \end{pmatrix}$ ∴ Instantaneous Velocity (×2) = $\begin{pmatrix} 15.25 \\ 9 \\ 4 \end{pmatrix}$

iii) $\underline{r}(2.1) - \underline{r}(2) = \begin{pmatrix} 9.261 \\ 8.82 \\ 8.4 \end{pmatrix} - \begin{pmatrix} 8 \\ 8 \\ 8 \end{pmatrix} = \begin{pmatrix} 1.261 \\ 0.82 \\ 0.4 \end{pmatrix}$ ∴ Instantaneous Velocity (×10) = $\begin{pmatrix} 12.61 \\ 8.2 \\ 4 \end{pmatrix}$

iv) $\underline{r}(2.01) - \underline{r}(2) = \begin{pmatrix} 8.1206 \\ 8.0802 \\ 8.04 \end{pmatrix} - \begin{pmatrix} 8 \\ 8 \\ 8 \end{pmatrix} = \begin{pmatrix} 0.1206 \\ 0.0802 \\ 0.04 \end{pmatrix}$ ∴ Instantaneous Velocity (×100) = $\begin{pmatrix} 12.06 \\ 8.02 \\ 4 \end{pmatrix}$

It can be seen that the velocity is tending to $\begin{pmatrix} 12 \\ 8 \\ 4 \end{pmatrix}$ in the limit.

When differentiated we would expect $\begin{pmatrix} t^3 \\ 2t^2 \\ 4t \end{pmatrix}$ to be :-

d.c of t^3 : $\frac{dr}{dt} = 3t^2$

d.c of $2t^2$: $\frac{dr}{dt} = 4t$ ∴ $\frac{dy}{dx}$ or $\frac{dr}{dt} = \begin{pmatrix} 3t^2 \\ 4t \\ 4 \end{pmatrix}$ = Velocity.

d.c. of $4t$: $\frac{dr}{dt} = 4$

Thus when $t = 2$:- $\frac{dr}{dt} = \begin{pmatrix} 12 \\ 8 \\ 4 \end{pmatrix}$

Which is the answer we got in the limit. The conclusion is therefore that we can differentiate 3d vectors as well as 2d.

NOTE :- acceleration = $\frac{d^2r}{dt^2} = \begin{pmatrix} 6t \\ 4 \\ 0 \end{pmatrix}$

> This is some of the work done by Paul on 'Movement and Vectors'. He has successfully produced average velocity vectors, but it took him quite some time before he made the connection with differentation.

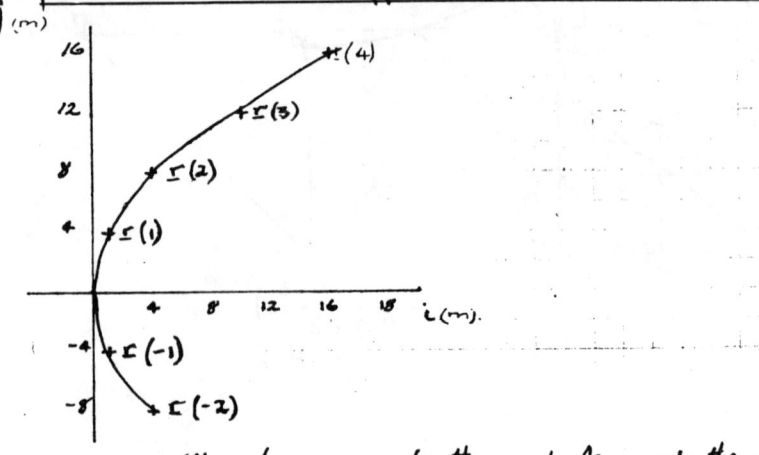

[Graph showing j(m) vs i(m) with points r(-2), r(-1), r(1), r(2), r(3), r(4) plotted on a parabolic curve]

After two seconds the particle is at the co-ordinates $4\underline{i} + 8\underline{j}$

* What is $\underline{r}(3) - \underline{r}(2)$?

$$= \binom{9}{12} - \binom{4}{8}$$

$$\underline{r}(3) - \underline{r}(2) = \binom{5}{4}$$ This is the direction vector which the particle travel

* $\underline{r}(4) - \underline{r}(2)$ * $\underline{r}(5) - \underline{r}(4)$ * $\underline{r}(4 \cdot 5) - \underline{r}(4)$

$= \binom{16}{16} - \binom{4}{8} = \binom{12}{8}$ $= \binom{9}{2}$ $= \binom{4 \cdot 25}{2}$

Av. Vel. $= \binom{6}{4}$ Av. Vel. $= \binom{9}{2}$ Av. Vel. $= \binom{8 \cdot 5}{4}$

* $\underline{r}(4 \cdot 1) - \underline{r}(4)$ * $\underline{r}(4 \cdot 01) - \underline{r}(4)$ * $\underline{r}(4 \cdot 001) - \underline{r}(4)$

$= \binom{0 \cdot 81}{0 \cdot 4}$ $= \binom{0 \cdot 0801}{0 \cdot 04}$ $= \binom{0 \cdot 008001}{0 \cdot 004}$

Av. Vel. $= \binom{8 \cdot 1}{4}$ Av. Vel. $= \binom{8 \cdot 01}{4}$ Av. Vel. $= \binom{8 \cdot 001}{4}$

* At infinity (in the limit), Av. Vel. $= \binom{8}{4}$

Mag $= \sqrt{8^2 + 4^2} = \sqrt{80} \approx 9 \, ms^{-1}$ $\underline{r} = \binom{t^2}{4t}$ $x = t^2$ $\frac{dx}{dt} = 2t$

direction $= \tan^{-1}(4/8) = \tan^{-1}(1/2)$ $y = 4t$ $\frac{dy}{dt} = 4$

$= 26 \cdot 57°$

$\therefore \underline{v} = \binom{2t}{4}$

$t = 4$ $\therefore \underline{v}(4) = \binom{8}{4}$

$\underline{v}(5) = \binom{10}{4}$

> After his earlier experience Rob has efficiently and confidently differentiated the position vector to obtain expressions for the velocity and acceleration vectors of a particle following a circular path.

Robert Worthington

Circular Motion

A particle P is moving round a circle of radius 6m with an angular velocity of 2 rads/sec.

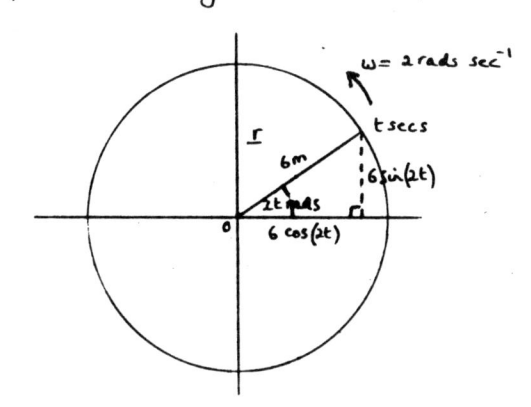

when time is 1 sec, angle = 2 rads
when time is t sec, angle = 2t rads

polar coordinates of $\underline{r}$ = $(6, 2t)$ = $6\cos\theta + 6\sin\theta$
$$= 6\cos(2t) + 6\sin(2t)$$

$\Rightarrow$ vector $\underline{r}(t) = \begin{pmatrix} 6\cos(2t) \\ 6\sin(2t) \end{pmatrix}$

differentiating $\underline{v}(t) = \begin{pmatrix} -12\sin(2t) \\ 12\cos(2t) \end{pmatrix}$

$\underline{a}(t) = \begin{pmatrix} -24\cos(2t) \\ -24\sin(2t) \end{pmatrix}$

Generalizing to a circle of radius R metres and angular velocity ω rads sec^{-1}

$$\underline{r}(t) = \begin{pmatrix} R\cos(\omega t) \\ R\sin(\omega t) \end{pmatrix} = R \cdot \begin{pmatrix} \cos(\omega t) \\ \sin(\omega t) \end{pmatrix}$$

Tangential to the circle

$$\underline{v}(t) = \begin{pmatrix} -\omega R \sin(\omega t) \\ \omega R \cos(\omega t) \end{pmatrix} = R\omega \cdot \begin{pmatrix} -\sin(\omega t) \\ \cos(\omega t) \end{pmatrix}$$

acting towards the circle's centre
$\perp$ to velocity

$$\underline{a}(t) = \begin{pmatrix} -\omega^2 R \cos(\omega t) \\ -\omega^2 R \sin(\omega t) \end{pmatrix} = -\omega^2 R \cdot \begin{pmatrix} \cos(\omega t) \\ \sin(\omega t) \end{pmatrix}$$

Chapter 4   Complex Numbers

4.1   Set of points in the Argand diagram

4.2   Complex number transformations

4.3   Sets of points and transformations

4.4   The transformation $z \rightarrow \frac{1}{z}$

4.5   More about $z \rightarrow \frac{1}{z}$

4.1 Set of points in the Argand diagram

A Sketch (or draw) the following curves:

$$x^2 + y^2 = 9$$
$$(x-2)^2 + y^2 = 9$$
$$x^2 + (y-4)^2 = 9$$
$$(x-2)^2 + (y-4)^2 = 9$$

What will $(x-a)^2 + (y-b)^2 = r^2$ look like?

B If $z = x + iy$ express the following in terms of x and y:

(1) $|z| = 3$ (2) $|z-2| = 3$ (3) $|z-4i| = 3$
(4) $|z-(2+4i)| = 3$ (5) $|z-(a+bi)| = 3$

Sketch the curves you get. Comments?

C Now use what you have discovered to sketch these quickly.

(1) $|z| = 4$ (2) $|z-2| = 2$ (3) $|z-4i| = 5$
(4) $|z-(2+4i)| = 2$ (5) $|z-(3-7i)| = 2$
(6) $|z+5| = 2$ (7) $|z+3i| = 3$
(8) $|z+6-2i| = 1$ (9) $|z-(a+bi)| = 2$
(10) $|z+a-bi| = 3$ (11) $|z-a+bi| = r$

Complex number transformations 4.2

A   Sketch the curve $|z| = 2$.

What happens to it under each of the following transformations?

$z \rightarrow z - 1$

$z \rightarrow z + 1$

$z \rightarrow z + i$

$z \rightarrow z - i$

$z \rightarrow z + 2 - 3i$

$z \rightarrow z - 2 + 5i$

$z \rightarrow 2z$

$z \rightarrow -z$

$z \rightarrow \dfrac{z}{3}$

$z \rightarrow 3z - 1$

$z \rightarrow 4 - 2z$

B   Repeat the above for the curve $|z - 2| = 2$

4.3 Sets of points and transformations

A Draw $\arg(z) = \frac{\pi}{4}$

Use it to sketch

(1) $\arg(z-1) = \frac{\pi}{4}$ (2) $\arg(z-i) = \frac{\pi}{4}$

(3) $\arg[z-(1+i)] = \frac{\pi}{4}$

(4) $\arg[z-(2-3i)] = \frac{\pi}{4}$

(5) $\arg[z-(a+bi)] = \frac{\pi}{4}$

B Now sketch these:

(1) $\arg(z-3) = \frac{\pi}{6}$

(2) $\arg(z-2i) = \frac{\pi}{2}$

(3) $\arg[z-(2+3i)] = \frac{\pi}{2}$

(4) $\arg(z+3-2i) = \frac{5\pi}{6}$

(5) $\arg(z-2+3i) = -\frac{3\pi}{4}$

(6) $\arg(z+2+5i) = -\frac{\pi}{2}$

(7) $\arg(z+a-bi) = -\frac{2\pi}{3}$

(8) $\arg(z-a+bi) = \frac{3\pi}{4}$

(9) $\arg(z+a+bi) = \theta$

The transformation $z \to \frac{1}{z}$

This sheet is about the transformation
$$z \to \frac{1}{z}$$

A Choose one of the lines
$$x=1\ ;\ x=\tfrac{3}{4}\ ;\ x=\tfrac{1}{2}\ ;\ x=\tfrac{1}{4}\ ;\ x=\tfrac{1}{3}$$
If you are working in a group try to be sure that somebody uses each line.

By plotting the images of selected points on the line on graph paper, find the image of your line under $z \to \frac{1}{z}$.

B Repeat A for one of the lines
$$y=1\ ;\ y=\tfrac{3}{4}\ ;\ y=\tfrac{1}{2}\ ;\ y=\tfrac{1}{4}\ ;\ y=\tfrac{1}{3}$$

C Plot $|z-a|=a$ accurately for small integer a of your choice. Find its image under $z \to \frac{1}{z}$ by selecting points and finding their images.

D Plot $|z-1-i|=1$ accurately. Find its image under $z \to \frac{1}{z}$ by plotting the images of selected points.

E Find the image of $y=mx$ for a selected value of m, under the mapping $z \to \frac{1}{z}$.

4.5 More about $z \to \frac{1}{z}$

A Look at the line you transformed in part A of the previous sheet.

Find a way to write down the general point on the line.

Find the image of the point under $z \to \frac{1}{z}$

Does the result confirm your previous answer?

B Repeat part A for the line you used in part B of the previous sheet.

C Use this diagram to help you find the image of

$$|z - a| = a \quad \text{under} \quad z \to \frac{1}{z}$$

D Find a way to represent the general point on $|z - 1 - i| = 1$

Hence confirm your results from the previous sheet.

E Write down the general point on $y = mx$, using your selected value of m from the last sheet.

Hence confirm your results from the previous sheet.

Comments on Complex Number Sheets

The intention of these five sheets is to make students comfortable with the Argand diagram. The first sheet builds up the standard work on circles in the complex plane.

The second sheet explores the geometrical effect of simple functions on a circle. This also gives useful practice in simple calculations.

The third sheet moves on to sets of points specified in terms of the argument of a complex number. This builds the ideas of sheet 2 into the work on arguments.

The fourth and fifth sheets are concerned with the transformation
$z \rightarrow \frac{1}{z}$

It has the property that:

i) Circles through the origin are transformed into straight lines through the origin and vice versa.

ii) Straight lines through the origin are reflected in the y axis.

iii) Circles not through the origin are transformed into other circles not through the origin.

Sheet 4 is to be done entirely graphically and so involves many calculations of the form:

$$\frac{1}{0.7 + 0.8i}$$

giving useful practice. When students worked in pairs on different lines it led to an awareness of the general result stated above. The comparison of notes between different pairs allowed general hypotheses to be formed about the centres of the image circles.

The detailed calculations seemed to lead to an awareness of the self-inverse nature of the transformation.

An interesting discussion arose from the fact that many people believed that the circle in D only had an image consisting of a quarter circle in which $|Z| < 1$ but one of the quieter members of the group explained decisively how stupid they were to imagine that by taking points for which $|Z| > 1$ that they could be getting all the possible points, since by taking images of points with $|Z| < 1$, they would get the missing $|Z| > 1$ parts of the circle.

Sheet 5 leads students to confirm the results by repeating the work of sheet 4.

Chapter 5   **Calculus**

5.1   Transforming derivatives

5.2   Transforming integrals

5.3   An area function for a trig curve

5.4   The differential equation $\frac{dy}{dx} = y$ step functions

5.5   The differential equation $\frac{dy}{dx} = y$ solution in series 1

5.6   The differential equation $\frac{dy}{dx} = y$ solution in series 2

5.7   The differential equation $\frac{dy}{dx} = \frac{1}{x}$ 1

5.8   The differential equation $\frac{dy}{dx} = \frac{1}{x}$ 2

5.9   Integral equations

5.10   Integral equations continued

5.1 Transforming Derivatives

A The graphs of the functions $x \to f(x)$ and $x \to Af(ax+b)+B$ are related in ways which have been investigated on the sketching functions sheets.

Now investigate what happens to the derivative when the function is transformed in this way.

Use sketch graphs (and computer graph programs if available) to investigate the relationship between the derivative of $f(x) = \sin x$ and the derivatives of $f(x) = \sin x + 1$, $f(x) = \sin x + 3$, $f(x) = \sin x + B$
$f(x) = \sin(x+1)$, $f(x) = \sin(x+3)$, $f(x) = \sin(x+B)$
$f(x) = \sin 2x$, $f(x) = \sin \frac{1}{2} x$, $f(x) = \sin 3x$
$f(x) = \sin ax$, $f(x) = 2\sin x$ ....

Write up any rules you find with an explanation of why they work.

B Test your rules on suitable transformations of
(1) $f(x) = \frac{1}{x^2}$ (2) $f(x) = (x+2)(x-1)$ (3) $f(x) = x^3 + x^2$

Transforming integrals 5.2

A The graphs of the functions $x \to f(x)$ and $x \to Af(ax+b) + B$ are related in ways which have been investigated on the sketching functions sheets.

Now investigate what happens to the integral when the function is transformed in this way.

Use sketch graphs to investigate the relationship between the the area function of $f(x) = \sin x$, ie $A(x) = \int_0^x \sin t \, dt$.

Continue with the area functions for the following: $f(x) = 1 + \sin x$, $f(x) = 3 + \sin x$, $f(x) = B + \sin x$, $f(x) = \sin(x+1)$, $f(x) = \sin(x-3)$, $f(x) = \sin(x+B)$, $f(x) = \sin 2x$, $f(x) = \sin \frac{1}{2} x$, $f(x) = \sin 3x$, $f(x) = \sin ax$, $f(x) = 2 \sin x$ etc.....

Write up any rules you find with an explanation of why they work.

B Test your rules on suitable transformations of

(1) $f(x) = \dfrac{1}{x^2}$ (2) $f(x) = (x+2)(x-1)$

(3) $f(x) = x^3 + x^2$

5.3 An area function for a trig curve

On graph paper, draw $y = \cos x$ for $0 \leq x \leq 90°$

use a scale 1 cm for 20° on the x axis
             1 cm for 0.2 on the y axis.

$A(x)$ is defined as follows:

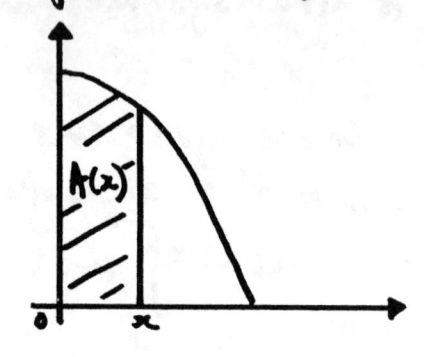

[diagram showing shaded area $A(x)$ under curve from 0 to x]

Complete the following table:

x	0	10	20	30	40	50	60	70	80	90
$A(x)$	0									57.1

Now plot $A(x)$ on a new piece of graph paper.

Suppose that we want to transform this graph so that the range is from 0 to 1 instead of from 0 to 57.1. How would we do it evenly?

Carry out this transformation. Do you recognise the graph of the new $A(x)$?

If not work out $A(100)$ etc.

The differential equation $\frac{dy}{dx}=y$ step functions 5.4

Investigate solutions to the differential equation
$$\frac{dy}{dx} = y$$

initially by using a compass needle diagram and then making use of the Euler step method.

For the solution curve passing through $(0,1)$ can you find better and better approximations to the value of y when $x=1$. Try to find a quicker method than by working through 100 or 1000 or 10000 steps! What happens if the number of steps is allowed to increase without limit?

If the solution curve through $(0,1)$ is represented by $y = f(x)$, you have just found, approximately, the value of $f(1)$.
What is $f(2)$, $f(3)$ ... $f(x)$? What about $f(0)$, $f(-1)$?
What about solution curves through $(0, y_0)$?
What about $\frac{dy}{dx} = ky$?

5.5 The differential equation $\frac{dy}{dx} = y$ solution in series 1

Suppose that we could find a solution of

$$\frac{dy}{dx} = y$$

What if the solution were a polynomial? It would look like

$$y = a_0 + a_1 x + a_2 x^2 + a_3 x^3 + a_4 x^4 \ldots$$

If the solution passes through $(0, 1)$ what is the value of a_0?

Differentiate the polynomial with respect to x. What is the value of the gradient at $(0,1)$ according to the differential equation and according to the differentiated polynomial? What is a_1?

By repeatedly differentiating can you find $a_2, a_3, a_4 \ldots a_n$?

Write out a series solution to the differential equation! Evaluate this for $x = 1$.

Does it agree with the results obtained on the previous sheet?

5.6 The differential equation — solution in series 2

- Use a suitable computer program to compare graphs of various truncations of the series produced on the previous sheet.

- This technique produces what is called the MACLAURIN SERIES for the function. It relies on working around $x=0$.

  Attempt to produce a general formula for a Maclaurin series for $f(x)$ a general differentiable function, by writing the coefficients in terms of $f(0)$, $f'(0)$, $f''(0)$ ......

When exploring a new operation, I find it important to look at its inverse at the same time. So, despite my decision to separate differentiation and integration into two dinstinct ideas and later to look at the link between them, my students looked at inverse differentiation when first being introduced to differentiation. Consequently they had been exposed to differential equations, like $\frac{dy}{dx} = x^2$ and they had used 'compass needle diagrams' to explore them quite early in the course.

They had also used a numerical step method, Euler's method, to obtain approximate values of the function. 'Teaching Calculus' by Shuard and Neill contains details of these and chapter 11 of 'Advanced Mathematics Book 2' by Perkins and Perkins also deals with them. So being asked to work on sheet 5.4 presented no great difficulty. 'Teaching Calculus' contains a detailed description of what happens as the number of steps, n, is increased and the further development of the approach to produce an exponential franction.

I felt that it was worth using the same technique to look at $\frac{dy}{dx} = \frac{1}{x}$, before performing the usual tricks.

Despite series expansions not actually being in our syllabus most of my students worked on sheets 5.5 and 5.6 and produced a series solution to $\frac{dy}{dx} = y$ and hence the expansion of 2^x.

SOME STUDENTS' WORK ON THE DIFFERENTIAL EQUATION $\frac{dy}{dx} = y$

A.

$$\frac{dy}{dx} = y$$

but $y = a_0 + a_1 x + a_2 x^2 + a_3 x^3 + a_4 x^4 + \ldots$

If the solution curve passes through $(0, 1)$

$y = 1$ when $x = 0$

$\therefore y = a_0 + \underbrace{(a_1 \cdot 0) + (a_2 \cdot 0^2)}_{=0} \ldots$

$\therefore y = a_0$

$1 = 1$

$a_0 = 1$

gradient at $(0, 1)$ is:—

$$\frac{dy}{dx} = a_1 + \underbrace{2a_2 x + 3a_3 x^2 \ldots}_{=0}$$

$y = \frac{dy}{dx} = 1 + 0 = 1 \quad \therefore a_1 = 1$

$$\frac{d^2 y}{dx^2} = 2a_2 + \underbrace{6a_3 x + 12 a_4 x^2 + 20 a_5 x^3}_{=0}$$

$y = \frac{d^2 y}{dx^2} = 1 = 2a_2$

$\therefore a_2 = \frac{1}{2}$

$$\frac{d^3 y}{dx^3} = 6a_3 + 24 a_4 x + 60 a_5 x$$

$y = \frac{d^3 y}{dx^3} = 1 = 6a_3$

$a_3 = \frac{1}{6}$

$$\frac{d^4 y}{dx^4} = 24 a_4 + 120 a_5 x$$

$y = \frac{d^4 y}{dx^4} = 1 = 24 a_4$

$a_4 = \frac{1}{24}$

In general for a_n :—

For $a_2 = \dfrac{1}{2 \times 1} = \dfrac{1}{2}$

For $a_3 = \dfrac{1}{3 \times 2 \times 1} = \dfrac{1}{6}$

For $a_4 = \dfrac{1}{4 \times 3 \times 2 \times 1} = \dfrac{1}{24}$

For $a_n = \dfrac{1}{n!}$ (MACLAURIN SERIES)

$$\sum_{r=1}^{\infty} r = \dfrac{1}{n!}$$

$y = a_0 + a_1 x + a_2 x^2 + a_3 x^3 + a_4 x^4 + a_5 x^5 \ldots$
$y = 1 + 1 + \dfrac{1}{2} + \dfrac{1}{6} + \dfrac{1}{24} + \dfrac{1}{120} \ldots$
$y = 2.72$
y tends to 2.72

On these two pages Brian has explored the implications of attempting to fit a polynomial to a solution curve of the differential equation $\dfrac{dy}{dx} = y$.

Series expansions for e^x, $\sin x$ etc are no longer on our A-level syllabus and he will not be asked to worry about the existence of such solutions or their convergence, other than very informally and intuitively.

B.

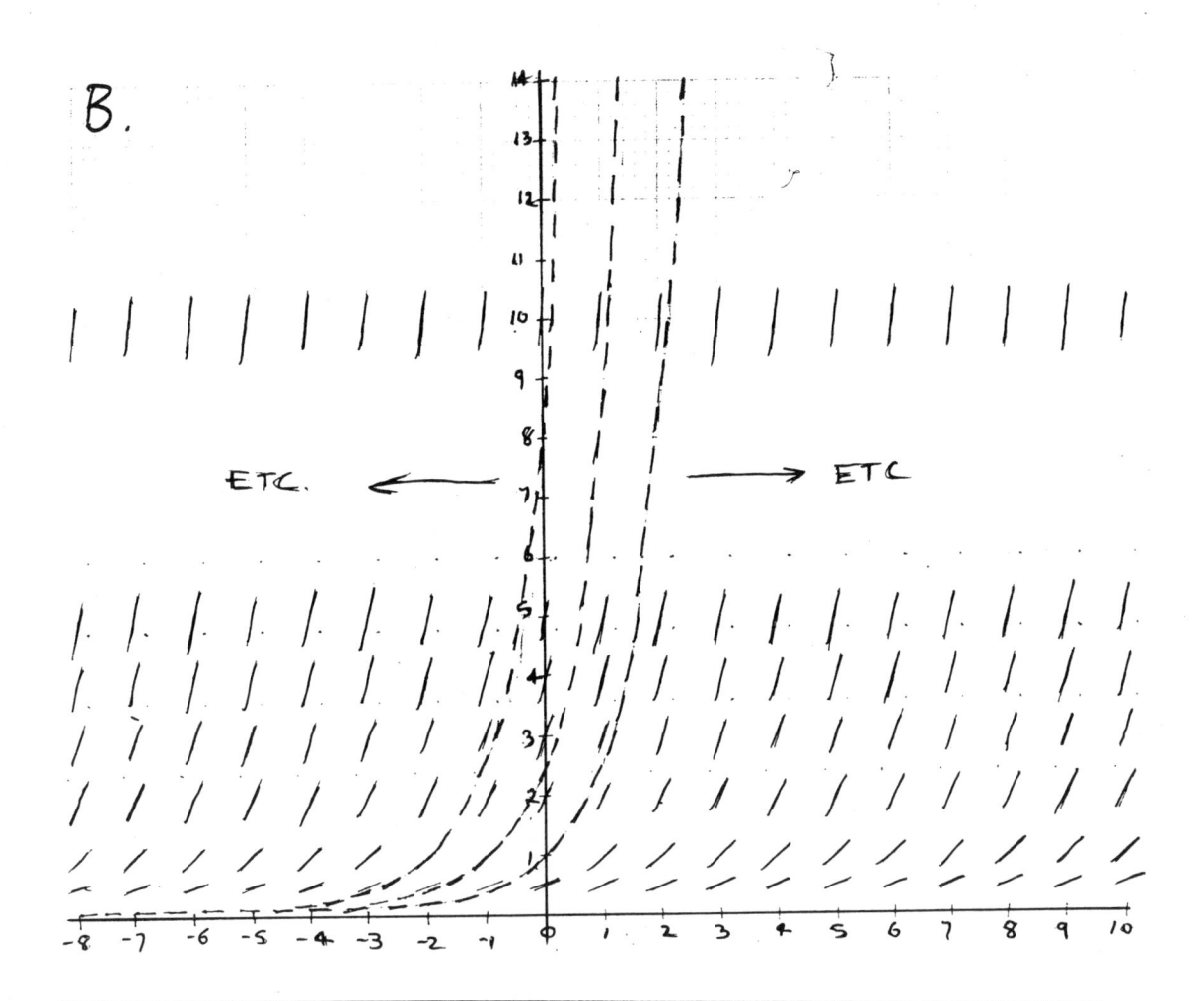

ETC. ←   → ETC

Here Phil has used a 'compass needle diagram' to help him to explore the differential equation $\frac{dy}{dx} = y$. He has sketched a few solution curves and is probably ready to produce some approximate numerical values using a step-by-step method. Investigation of a step solution will enable him to deduce all the properties of the exponential function. He will also need to look at the limit of $\left(1 + \frac{1}{n}\right)^n$ as $n \to \infty$.

5.7 The differential equation $\frac{dy}{dx} = \frac{1}{x}$ 1

Consider the differential equation

$$\frac{dy}{dx} = \frac{1}{x}$$

Sketch a compass needle diagram and some solution curves for the differential equation.

Choose the solution through $(1, 0)$ and apply a step-by-step method to find values of the solution for whole number values of x between 1 and 10.

Draw a graph of this function.

Investigate $f(a) + f(b)$ and $f(a) - f(b)$

What about solutions through $(1, 1), (1, 2)$?

What about solutions through $(-1, 0), (-1, 2)$?

The differential equation $\frac{dy}{dx} = \frac{1}{x}$ 2

Now we look for a solution of

$$\frac{dy}{dx} = \frac{1}{x} \quad \text{as a series.}$$

We shall look for a polynomial approximation to the curve through $(1, 0)$

The method we used for $\frac{dy}{dx} = y$ does not quite work! Why?

Translate the solution 1 unit to the left. What differential equation will have this as a solution?

Use the methods of sheet 5.5 to find the coefficients in $y = a_0 + a_1 x + a_2 x^2 \ldots$ by repeatedly differentiating!

Translate the polynomial to the right one unit!

Use a suitable computer program to compare graphs of $\ln(x)$ with the polynomial you have found to a few orders of approximation.

Integral equations

For what function f is this equation true?
$$\int_0^x f(t)\,dt = \frac{x^3}{3}$$

For what function f is this equation true?
$$\int_0^x f(t)\,dt = 1 - \cos x, \text{ where } x \text{ is in radians.}$$

If we have no easy method of writing down what f is we can explore the equation graphically and numerically.

What are the values of $\int_0^{0.1} f(t)\,dt$; $\int_{0.1}^{0.2} f(t)\,dt$ ...... ?

Complete a table like this one:

x-values (a, b)	Area between a and b
0, 0.1	
0.1, 0.2	
0.2, 0.3	
$\vdots$	

On graph paper draw rectangles to represent the area between 0 and 0.1
between 0.1 and 0.2 and so on...
How do you calculate the height of each rectangle?

continued on next sheet...

Integral equations continued

How can you estimate values of $f(0.1)$, $f(0.25)$?

How can you produce more accurate values?

Produce some approximate values for $f(0.1)$, $f(0.25)$ ... as accurately as your calculator will allow.

Use these to sketch a graph of $x \to f(x)$
Describe your method in detail.

Describe the method for a general equation
$$\int_a^x f(t)\,dt = F(x)$$ where F is some area function

Now try the method on these :-
$$\int_0^x f(t)\,dt = \tan x \qquad \int_0^x f(t)\,dt = \sin x$$
$$\int_0^x f(t)\,dt = \cos x \qquad \int_0^x f(t)\,dt = \ln x$$

and on some of your own

Not all of the above work. Why not?
Find some way of changing the equation so that it is soluble.

Integral Equations Teachers Notes

Integral equations like $\int_0^x f(t)dt = 1 - \cos x$ can be investigated, and approximate solutions found, using this method.

Graphically, this sort of diagram is produced;

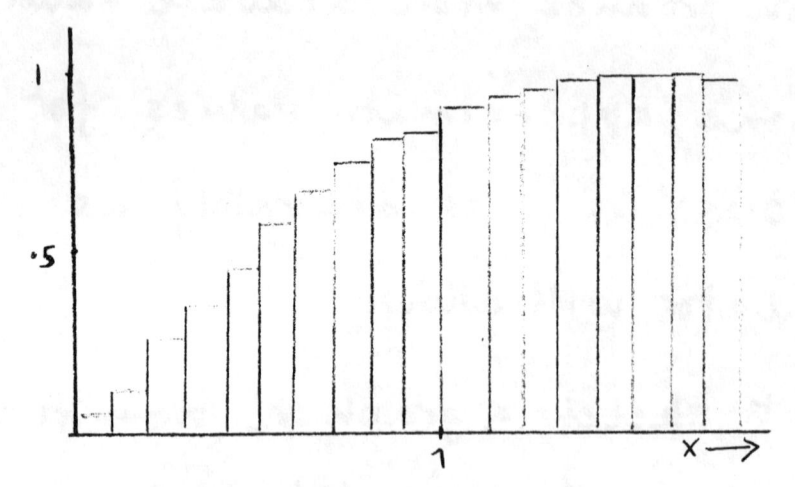

Exploring more accurate estimates of the value of f at particular points by decreasing the size of interval can provide the sort of experience needed to prepare students to look at $\frac{(x+h) - f(x)}{h}$

and to make the connection between integration, as an area of measure or area function and differentiation as a rate function.

It is worth asking students to explore other equations of the same sort, especially those involving functions which they have not yet dealt with. A randomly inverted equation is likely to be insoluble - or inconsistent!

Chapter 6   **Differences**

6.1   Differences 1

6.2   Differences 2

6.3   Differences 3

6.4   Using Differences 1

6.5   Using Differences 2

6.6   Using Differences 3

6.7   Using Differences 4

6.8   Using Differences 5

6.9   Program for Differences

6.10   Using the differences Program

6.1 Differences 1

The sequence of squares can be 'differenced' like this:

$$1 \quad 4 \quad 9 \quad 16 \quad 25 \quad 36 \quad 49 \quad 64$$
$$3 \quad 5 \quad 7 \quad 9 \quad 11 \quad 13 \quad 15$$
$$2 \quad 2 \quad 2 \quad 2 \quad 2 \quad 2$$

We reach a CONSTANT DIFFERENCE after two DIFFERENCE ROWS.

Study the difference patterns for each of these sequences:

$a_n = 2n^2$ $\quad a_n = 3n^2 \quad$ $a_n = n^2 + 2n$

$a_n = n^2 + 2n + 10$ $\quad a_n = n^3 \quad$ $a_n = 4n^3$

$a_n = 4n^3 + n^2$ $\quad a_n = 2n^3 - 3n^2 + 1$

$a_n = n^4$ $\quad a_n = n^5 \quad$ $a_n = n^4 + n^3 + n^2$

You may like to use a computer to generate the initial sequence!

Can you predict the constant difference?

Can you predict the number of difference rows?

Differences 2   6.2

If you are given a sequence of numbers, can you now determine the highest power term involved in it?

Test your ideas on these:

10, 15, 22, 31, 42, 55, 70, 87, 106, 127

-1, 2, 9, 20, 35, 54, 77, 104, 135, 170

$4\frac{1}{2}$, 11, $18\frac{1}{2}$, 27, $36\frac{1}{2}$, 47, $58\frac{1}{2}$, 71, $84\frac{1}{2}$, 99

7, 20, 45, 88, 155, 252, 385, 560, 783, 1060

3, 16, 53, 126, 247, 428, 681, 1018, 1451, 1992

4, 17, 80, 253, 620, 1289, 2392, 4085, 6548

148, 142, 132, 118, 100, 78, 52, 22, -12, -50

Can you think of any way to continue further in determining the sequence?

6.3 Differences 3

A   It would be useful to be able to find not just the highest power term of a sequence, but all its terms.

Perhaps we could just subtract the sequence given by the highest term ... ?

Can we get rid of fractions too ... ?

Experiment with the sequences of differences 1.

B   Apply the differencing methods to this sequence:

1, 1, 2, 3, 5, 8, 13, 21, 34, 55, 89

What do you think your results mean?

Using Differences 1 6.4

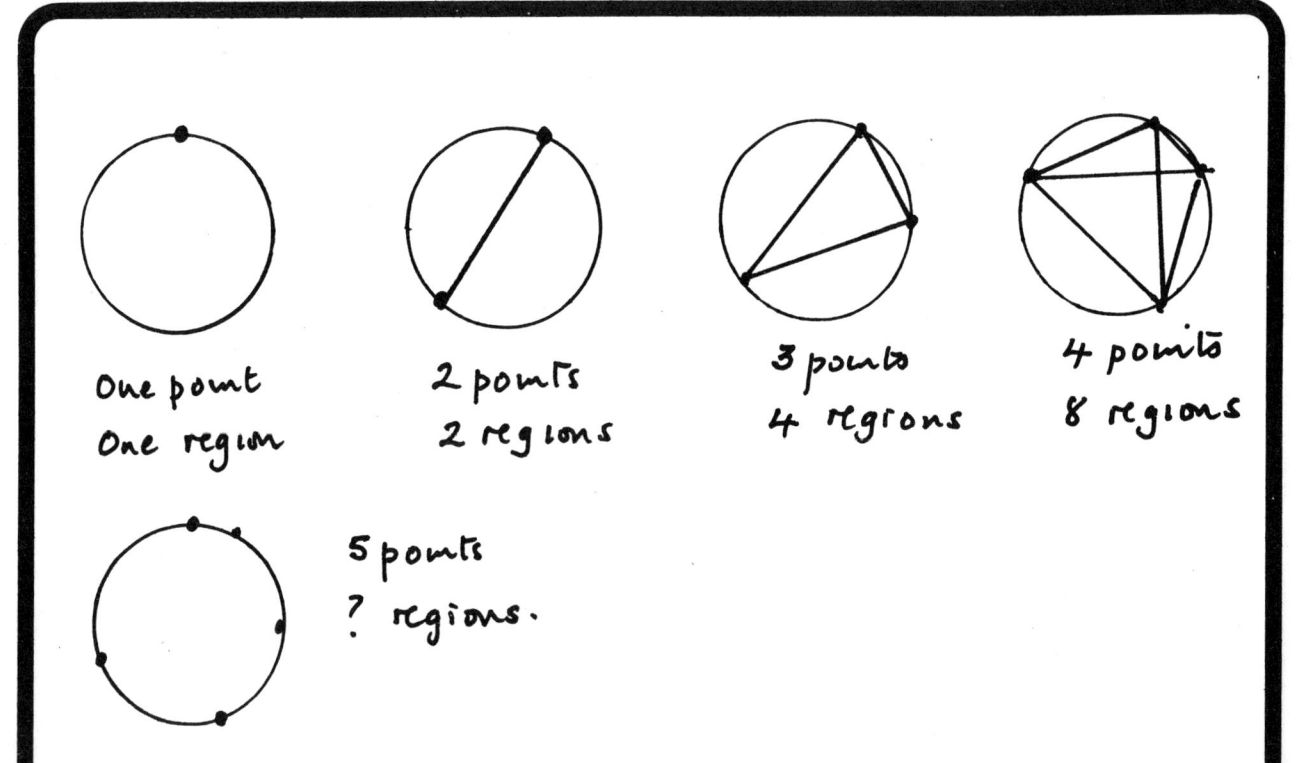

One point
One region

2 points
2 regions

3 points
4 regions

4 points
8 regions

5 points
? regions.

Continue the sequence up to 7 points.
DONT make assumptions.

Can you find a formula for the sequence using difference methods?

6.5 Using Differences 2

Look at this cube:

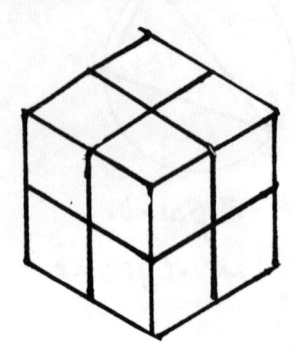

There are   8   1-cubes in it
            1   2-cubes in it
           ―――
            9   cubes all told.

Generalise to other sizes of cube.

Examine the sequence produced.

Can you find the general formula for it?

Using Differences 3

Look at this cube:

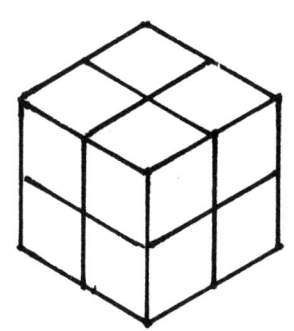

There are 12   (1×1×1) - cuboids
           8   (2×1×1) - cuboids
           6   (2×2×1) - cuboids
           1   (2×2×2) - cuboids
        ―――
        27   cuboids in all.

Generalise to other sizes of cube. Examine the sequence produced. Can you find a general formula for it?

Differences

My first year further maths group found that there were

$$1^2 + 2^2 + 3^2 + \ldots + n^2$$

squares on an n x n chessboard after much arguing and making of mistakes. At this stage they were unable to find a simple formula for $\sum r^2$.

They then worked through the differences sheets 1 and 2 as homework, over the period of about a week so that they were discussing their findings with me and with each other over this time.

They were then able to find out, by differences that

$$\sum_{r=1}^{n} r^2 = \frac{1}{6}n(n+1)(2n+1)$$

which worked for all the cases they tried. This seemed to be a suitable point to talk about proof, so I introduced the idea of induction, talking in detail about the strategy before doing any writing. When we did come to write down a proof of their formula they still found the algebraic manipulation difficult but because they understood and were in sympathy with the general approach they were able to keep a grip on the necessary steps.

Using Differences 4 6.7

A If we have a 3×3 pin-board:

. . .

. . .

. . .

We can make 6 squares on it if we allow 'slanting' squares:

4 squares   1 square   1 square. Total 6

Generalise ....

B Generalise to an m×n board.

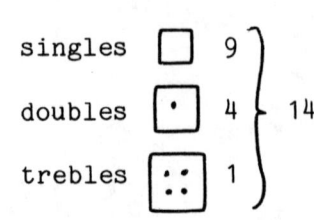

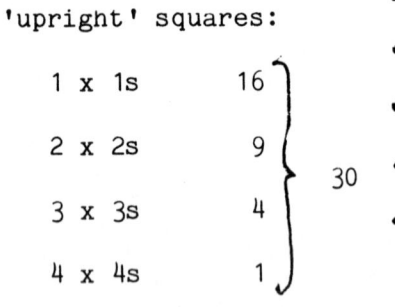

Solution to 'Using Differences 4' Part A

1 x 1 – pinboard: 0 squares

2 x 2 – pinboard: 1 square

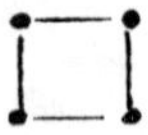

3 x 3 – pinboard: 6 squares

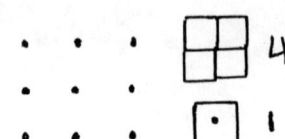

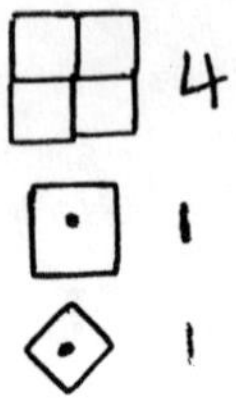

              4
              1
              1

4 x 4 – pinboard: 20 squares

 'upright' squares:

  singles   9 ⎫
  doubles   4 ⎬ 14
  trebles   1 ⎭

 'diagonal' squares:

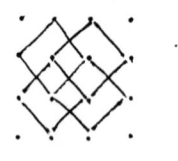

  side $\binom{1}{1}$ 4 ⎫
  side $\binom{1}{2}$ 1 ⎬ 6
  side $\binom{2}{1}$ 1 ⎭

5 x 5 – pinboard: 50 squares

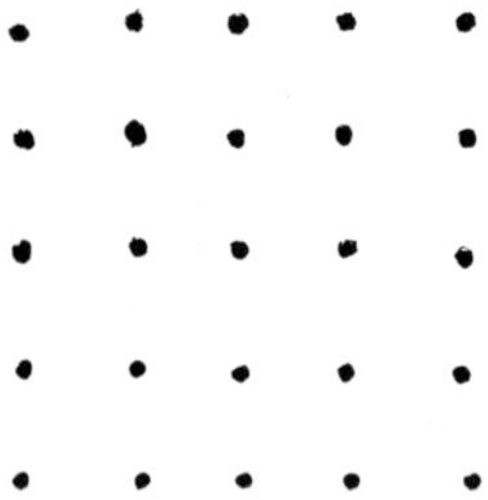

 'upright' squares:

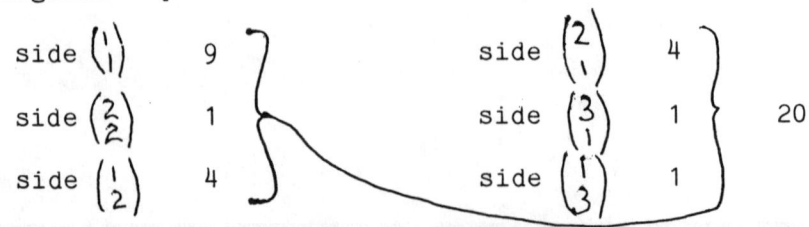

  1 x 1s   16 ⎫
  2 x 2s   9 ⎪
        ⎬ 30
  3 x 3s   4 ⎪
  4 x 4s   1 ⎭

 'diagonal' squares:

  side $\binom{1}{1}$ 9   side $\binom{2}{1}$ 4 ⎫
  side $\binom{2}{2}$ 1   side $\binom{3}{1}$ 1 ⎬ 20
  side $\binom{1}{2}$ 4   side $\binom{1}{3}$ 1 ⎭

6 x 6 – pinboard:

. . . . . .
. . . . . .
. . . . . .
. . . . . .
. . . . . .
. . . . . .

105 squares

'upright' squares:

25 + 16 + 9 + 4 + 1 = 55

'diagonal' squares:

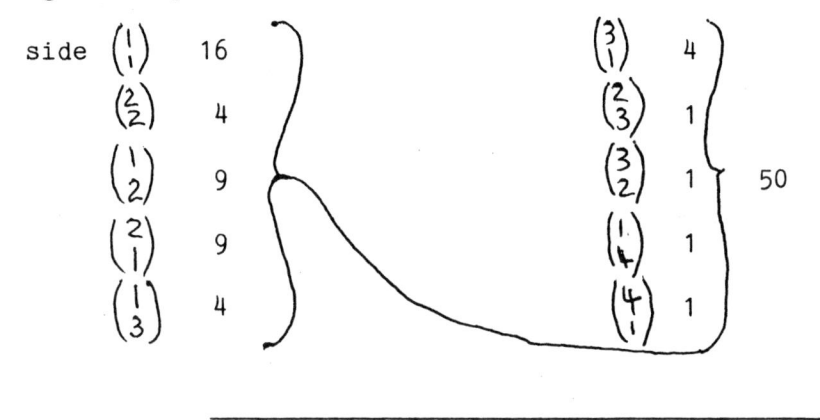

side $\binom{1}{1}$ 16
$\binom{2}{2}$ 4
$\binom{1}{2}$ 9
$\binom{2}{1}$ 9
$\binom{1}{3}$ 4

$\binom{3}{1}$ 4
$\binom{2}{3}$ 1
$\binom{3}{2}$ 1
$\binom{1}{4}$ 1
$\binom{4}{1}$ 1

} 50

We now have the following sequence for the number of squares on an n x n pinboard:

0, 1, 6, 20, 50, 105, ...

Next, this sequence is tabulated and successive differences calculated:

n	1	2	3	4	5	6
No of squares	0	1	6	20	50	105
1st differences	1	5	14	30	55	
2nd differences		4	9	16	25	
3rd differences			5	7	9	
4th differences				2	2	

A constant difference of 2 appears on the 4th difference row so the formula is of the form:

$$S_n = \frac{1}{12} n^4 + Bn^3 + Cn^2 + Dn + E$$

The simplest procedure is now to multiply the sequence by 12:

n	1	2	3	4	5	6
$12S_n$	0	12	72	240	600	1260
n^4	1	16	81	256	625	1296
$12S_n - n^4$	-1	-4	-9	-16	-25	-36

Hence $\quad 12 S_n - n^4 = -n^2$

$\therefore \quad 12 S_n = n^4 - n^2$

$\quad\quad S_n = \dfrac{1}{12} n^2 (n^2 - 1)$

Using Differences 5   6.8

A Count the bricks used in building each of these shapes.

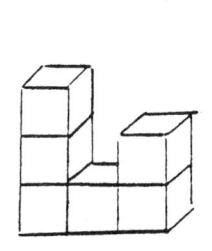

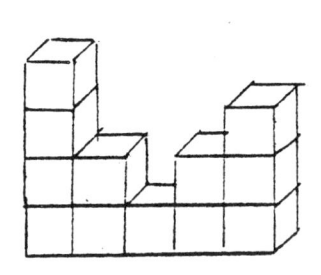

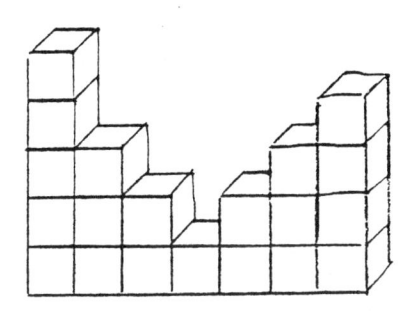

Continue the sequence for 2 more terms. Generalise . . . .

B Do the same for these 2 sequences.

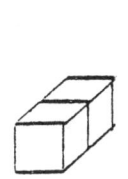

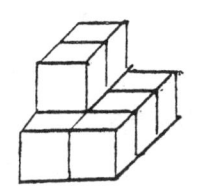

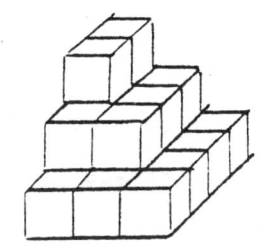

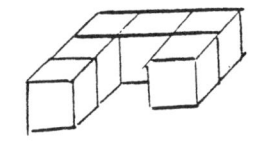

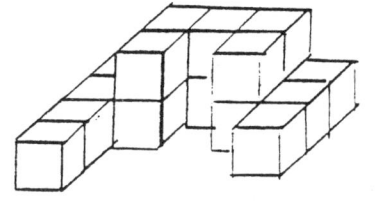

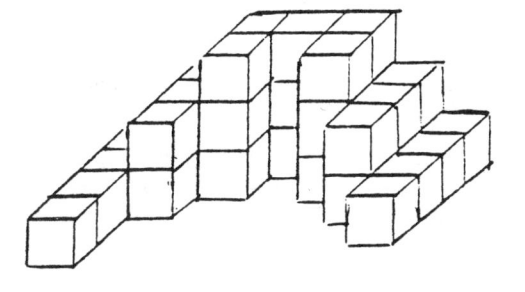

6.9 Program for Differences

```
0 CLS
5 PRINT"THE PROGRAM WILL TELL YOU WHETHER IT IS EXTRAPOLATING THE SEQUENCE OR NoT."
6 PRINT"THE PROGRAM WILL INFORM YOU IF THE       PROCEDURE FAILS."
7 PRINT
9 PRINT"FIRST INPUT THE NUMBER OF TERMS YOU ARE GOING TO GIVE IT."
10 INPUTN:M=N-1:DIMA(N+2,M+2):DIMB(M)
11 PRINT
12 DIMC(N+2,M+2)
14 G=0
17 PRINT"NOW INPUT THE TERMS IN ORDER."
19 PRINT
20 FORJ=0TOM:INPUTY:A(0,J)=Y:NEXT
21 FORJ=0TOM:C(0,J)=A(0,J):NEXT
22 R=A(0,0)
23 C=0:E=0
24 G$=""
30 C=C+1:V=M-C+1
40 FOR J=0TOV-1:U=C-1:Q=J+1:A(C,J)=A(U,Q)-A(U,J):NEXT
41 IF V= 1 THEN 500
50 GOSUB250
60 IF T = 0 THEN 70
62 GOTO 30
70 C1=C1+1:IF C1 = 1 THEN E1 =C-1
71 C= C-1
80 F=1:FORJ=1TOC:F=F*J:NEXT
86 D=A(C,0)/F
87 IF D= INT(D) THEN 94
88 GOSUB 1000
89 GOSUB 1050
90 IF G>1 THEN 1200
92 Q=1
94 B(C)=D:E=E+1
100 G$=G$+"+"+STR$(D)+"X^"+STR$(C)
104 GOTO 110
110 FORJ=0TOM:H=((J+1)^C)*D:A(0,J)=A(0,J)-H:NEXT
120 GOSUB250
132 IF T= 0 THEN PRINT"REQUIRED FORMULA IS:-":PRINTC(0,0)/R;"*Y=";G$:GOTO300
140 C=0:GOTO30
250 T=0:FORJ=0TOV:T=T+ABS(A(C,J)):NEXT:RETURN
300 PRINT
301 IF G=0 THEN E = E+1
307 PRINT"INPUT THE END POINTS OF A RANGE OF       VALUES TO BE EVALUATED,AND THE STEP.     THEN INPUT 1 OR THE DIVISOR NEEDED TO    RETURN TO THE ORIGINAL SEQUENCE"
308 PRINT
310 INPUTX,Y,Z,W
320 FOR L = X TO Y STEP Z: K=0
330 FOR J = 0 TO E1:K=K+(L^J)*B(J):NEXTJ
340 PRINT L;"   ";K/W:NEXT
380 END
500 PRINT
501 H=H+1
502 IF H>1 THEN PRINT"SEQUENCE BEYOND SCOPE OF PROGRAM":STOP
503 PRINT"THIS WILL BE A POLYNOMIAL APPROXIMATION "
510 A(C,1)=A(C,0):A(C,2)=A(C,0)
520 FOR I = C-1 TO 0 STEP -1
530 J=C-I+1:A(I,J)=A(I+1,J-1)+A(I,J-1):A(I,J+1)=A(I,J)+A(I+1,J)
539 NEXT I
540 M =M+2:E=E-1
545 PRINT"EXTENDEDSEQUENCE IS ":FOR I = 0TO M: PRINT A(0,I):NEXT I
550 GOTO 21
1000 FOR I =2 TO 1000
1001 Q=D*I
1002 IF Q-INT(Q) <0.00000001 THEN Q=I:D=D*I :I=1000
1003 NEXT I
1004 RETURN
1050 G=G+1
1100 FOR I= 0 TO M:A(0,I)=A(0,I)*Q:C(0,I)=C(0,I)*Q:NEXT I
1101 RETURN
1200 FOR I = 0 TO M:A(0,I)=C(0,I):NEXT I
1205 GOTO 23
```

Using the differences Program 6.10

Load the 'differences' program into a computer.

Run it for each of the following sequences in turn:

(1)   2, 9, 28, 65, 126, 217

(2)   1, 3, 10, 15, 21, 28, 36

(3)   1, 1, 2, 3, 5, 8, 13

(4)   3, 3, 5, 4, 4, 3

(5)   3, 3, 5, 4, 4, 3, 5

Comment on the variety in the response.

Use the program to check your own results for the difference sheets.

Chapter 7 **Calculator Work**

7.1   Calculator Accuracy 1
7.2   Calculator Accuracy 2
7.3   Working with Polynomials
7.4   Calculating $\cos\theta$

Notes on Chapter 7

Calculators are very useful tools at any age, but particularly in the sixth form. But how much do sixth formers really know about their own calculators?

The activities in Calculator Accuracy 1 and 2 are intended to be used to generate discussion within a class lesson. The aim of the work is to see how much the students know about their calculators and those of their neighbours. For example, the realisation that calculator answers cannot be taken at face value. It is important to do this activity with others so that answers produced by different calculators can be compared.

A-level students should be able to suggest a variety of methods to look at 'hidden figures' so the teacher may prefer not to circulate the worksheet as such but use as many of the prompts as prove necessary. It could be useful to ensure that different calculators are available by asking students to bring in others from home or by using school ones as a supplement.

Working With Polynomials is a worksheet which was used in a tertiary college early in the A-level course when graphs of functions were being plotted. It provided a good activity to think about previous work on quadratics, produced discussion and did not seem to provide a threat to anyone. It seemed useful early in the course to discuss calculator uses and revealed an amazing variety of practice in feeder secondary schools. Calculators were seen to be important as an aid, which needed to be used carefully.

Calculating $\text{Cos}\theta$ was used after series expansions of trigonometric functions had been done and provides a use for these formulae. It proved useful to get the students to explain the method. It posed questions such as 'how small is small?' It answered questions such as 'how might caclulators work out trig. functions?'

Calculator Accuracy 1   7.1

Calculators often know more about an answer than we think.

These tasks are designed to encourage you to think about <u>your</u> calculator and the way it works.

For example, do you know how many significant figures are stored in your calculator? Does it round up? Does it round down?

Work with someone else who has a different calculator so that you can compare and discuss results.

- Follow this sequence, noting the answer at each stage:

    | 2 | ÷ | 3 | = | 1/x | 1/x |

- Now try this one, starting from a clear calculator.

    | 0.6666666 | 1/x | 1/x |

    (Feed in more sixes if your display will take it and try again)
    Can you explain what happens?

- Investigate for other fractions.

7.2 Calculator Accuracy 2

(This sheet follows on from Calculators 1)

- Now try [2] [÷] [3] [=] [−] [current display] [=]

  What do you learn here?
  Investigate for other fractions.

- Next try [2] [÷] [3] [=] (note this display)

  [×] [10] [=]

  [−] [original display] [=]

- Next try [2] [÷] [3] [=]

  [×] [100000] [=]

  [−] [66666] [=]

- Finally, on this theme, try

  [2] [÷] [3] [=] [×] [3] [=]

- What is the largest multiplication your calculator can cope with accurately? Select a multiplication a little too big for your calculator. Can you work it out accurately? Try other big numbers e.g
   123456789 × 987654321.

- $\frac{1}{17}$ is a recurring decimal with a period of 16 digits. Use appropriate methods from the above to evaluate $\frac{1}{17}$ accurately using your calculator. Try for other fractions.

Working with Polynomials

How would you calculate
$$f(x) = 2x^2 + 4x + 5 \text{ for } x = 0, 1, 2 \ldots ?$$
Is it easier to use $f(x) = 2(x+1)^2 + 3$?
Is your algorithm still the best if x increases in steps of 0.01 say?

Work out $f(1) - f(0)$
$f(2) - f(1)$ etc

What do you notice? Why?
Construct a new algorithm for calculating
$f(1), f(2), f(3)$ etc.
Will it work backwards giving $f(-1)$ etc?
Try to general to all quadratics.

- Look at $f(1.5), f(2.5)$ etc and develop an algorithm for $f(0.5n)$.

- Develop an algorithm for calculating
$$f(x) = \frac{1}{x} - \frac{1}{x+3}$$ for successive integers.

- Try some cubics.
  e.g. What is the best way of calculating:
  $f(x) = x^3 + 3x^2 + 3x + 9$
  $f(x) = x^3 + 3x^2 + 11x + 9$
  $f(x) = x^3 + 12x^2 + 6x$

  Investigate algorithms for calculating
  $f(1), f(2)$ etc.

7.4   Calculating Cos θ

This method will calculate $\cos \theta$ ($0 \leq \theta \leq \frac{\pi}{2}$)

It uses both $\cos x \simeq 1 - \frac{x^2}{2}$ for small x

and $\cos 2x = 2\cos^2 x - 1$

To calculate $\cos \theta$, divide θ by 2 n times to give

$$\cos \frac{\theta}{2^n} \simeq 1 - \frac{1}{2}\left(\frac{\theta}{2^n}\right)^2$$

Then use the $\cos 2x$ formula n times to give an approximate answer for $\cos \theta$. Use your calculator to evaluate these.

- Investigate the appropriate values of n necessary to obtain a given accuracy.
- Is there a limit to accuracy with this method?
- Does it help to use $\cos x = 1 - \frac{1}{2}x^2 + \frac{1}{24}x^4$?
- What about using $\cos 3x = 4\cos^3 x - 3\cos x$?
  (or $\cos 4x = $?)

Adapt the method to find $\tan \theta$, $\log_e x$ and e^x.